REMARKABLE CREATURES

Epic Adventures in the Search for the Origins of Species

∾ SEAN B. CARROLL ∾

Quercus

First published in Great Britain in 2009 by Quercus
This paperback edition first published in 2009 by

Quercus
21 Bloomsbury Square
London
WC1A 2NS

Published by permission of Houghton Mifflin Harcourt

A CIP catalogue record for this book is available
from the British Library

ISBN 978 1 84916 072 8

Printed and bound in Great Britain by Clays Ltd, St Ives plc

10 9 8 7 6 5 4 3 2 1

REMARKABLE CREATURES

'Carroll has an agreeably conversational style, giving personal immediacy to what might so easily have been a mere progress-report on the quest for natural origins during the past 150 years . . . Offers a bracing tonic for those whose rational enjoyment of the natural universe currently seems in danger of being overwhelmed by the strident infantilism of creationists'   *Spectator*

'Leaves you with an overwhelming sense of awe and respect for the most remarkable of creatures, the men and women who searched for the origins of species and in doing so gave us a profound sense of place among life on Earth'   *Guardian*

'[Of] the books published to mark Darwin's anniversaries this year . . . best, by a narrow margin, is Sean Carroll's *Remarkable Creatures*, which manages to combine a wide narrative sweep with wonderful details and superb writing'   *Financial Times*

'Sean Carroll is both a leading exponent of the modern search and a forceful public advocate for biological science as one of the glories of our culture . . . not yet another trip down the heritage nostalgia trail, but the celebration of a vital tradition that began with Darwin and becomes stronger with every passing year'   *Daily Mail*

'Delightful . . . [Sean Carroll] understands and conveys the adventure of scientific discovery, the passionate desire to know, that united them all'   *Sunday Telegraph*

ABOUT THE AUTHOR

Sean B. Carroll is an investigator at the Howard Hughes Medical Institute and professor of genetics at the University of Wisconsin-Madison. He is the author of *Endless Forms Most Beautiful: The New Science of Evo Devo* and *The Making of the Fittest: DNA and the Ultimate Forensic Record of Evolution*. He is one of the leading biologists of his generation, and his seminal scientific discoveries have been featured in *Time*, the *U.S. News & World Report*, *Discover*, the *New York Times*, and other publications.

www.seanbcarroll.com

FOR JAMIE, WILL, PATRICK, CHRIS, AND JOSH —
THE MOST REMARKABLE CREATURES IN MY WORLD

∾ Contents ∾

What is it that confers the noblest delight? What is that which swells a man's breast with pride above that which any experience can bring to him? Discovery! To know that you are walking where none others have walked; that you are beholding what human eye has not seen before; that you are breathing a virgin atmosphere. To give birth to an idea—to discover a great thought—an intellectual nugget, right under the dust of a field that many a brain-plow had gone over before . . . To be the first—that is the idea. To do something, say something, see something before anybody else—these are the things that confer a pleasure compared with which other pleasures are tame and commonplace . . . These are the men who have really lived—who have actually comprehended what pleasure is—who have crowded long lifetimes of ecstasy into a single moment.

— *Mark Twain*, Innocents Abroad *(1869)*

REMARKABLE
CREATURES

❧ PREFACE ❧
A Mixture of Spirit and Deed

NOT SO LONG AGO, most of the world was an unexplored wilderness. The animals, plants, and people that inhabited the lands beyond Europe were unknown, at least as far as the Western world was concerned. The rivers and jungles of the Amazon, the Badlands of Patagonia and of the American West, the tropical forests of Indonesia, the savannah and center of Africa, the vast interior of Central Asia, the polar regions, and the many chains of islands that dot the oceans were complete mysteries.

And, while our knowledge of the world's living inhabitants was slim, our grasp of our planet's past was nonexistent. Fossils had been known for millennia, but they were seen in the light of local mythologies about dragons and other imagined creatures, not in the light of natural science.

Our sense of the time scale of life on earth? Vague and vastly off the mark.

And our picture of our own species' history? A set of fantastic myths and fairy tales.

The explorations of the previously unseen parts of the world and the unearthing of the history of life and our origins are some of the greatest achievements in human history. This book tells the stories of some of the most dramatic adventures and important discoveries in two centuries of natural history—from the epic journeys of pioneering naturalists to the expeditions making headlines today—and how they inspired and have expanded one of the greatest ideas of modern science: evolution.

We will encounter many amazing creatures of the past and present, but the most remarkable creatures in these stories are the men and women. They are, without exception, remarkable people who have experienced and accomplished extraordinary things. They have lived the kinds of lives that Twain extolled—they walked where no others had walked, saw what no one else had seen, and thought what no one else had thought.

The people in these stories followed their dreams—to travel to far-away lands, to see wild and exotic places, to collect beautiful, rare, or strange animals, or to find the remains of extinct beasts or human ancestors. Very few started out with any notion of great achievement or fame. Several lacked formal education or training. Rather, they were driven by a passion to explore nature, and they were willing, sometimes eager, to take great risks to pursue their dreams. Many faced the perils of traveling long distances by sea. Some confronted the extreme climates of deserts, jungles, or the Arctic. Many left behind skeptical and anxious loved ones, and a few endured years of unimaginable loneliness.

Their triumphs were much more than survival and the collecting of specimens from around the world. A few pioneers, provoked by a riot of diversity beyond their wildest imaginations, were transformed from collectors into *scientists*. They posed and pondered the most fundamental questions about Nature. Their answers sparked a revolution that changed, profoundly and forever, our perception of the living world and our place within it.

Unlike their privileged countrymen back in the universities, churches, and drawing rooms of Europe, most of whom believed that the origins of living things was a matter outside the realm of natural science, these explorers asked not just what existed, but wondered how and why these creatures came to be. Unlike their teachers, who pursued a natural theology that interpreted everything in Nature as part of the design of a Creator—peaceful, harmonious, stable, and unchanging—this new cadre of naturalists discovered that Nature was, in fact, a dynamic and perpetual battleground in which creatures competed and struggled to survive, a war in which they either adapted and changed or were exterminated. Unlike their predecessors, who explained the distribution of living species in the world much as one would the instant and premeditated placement of pieces on a chessboard, these naturalists discovered that the world and the life it contained had a very long history that shaped where various plants and animals were found across the globe. And, unlike their contemporaries who viewed everything in Nature as being purposely created for man's benefit and domination, they rejected that conceit and placed humans within the animal kingdom, with our own earthly origins.

The torch of this revolution has been passing from generation to generation of scientists who have been walking, literally and figuratively, in the footsteps of these pioneers.

A mixture of spirit and deed

My goals in writing this book are to bring to life the pursuit and the pleasure of scientific discovery, at the same time capturing the significance of each advance for evolutionary science. The idea here is that science is far better understood and enjoyed, and made memorable, when we follow the bumpy roads of the scientists that led, eventually, to their achievements. It is not an original idea. Like the naturalists described here who trod in their predecessors' footsteps, I am following the lead of authors such as Paul de Kruif (*Microbe Hunters*) and C. W. Ceram (*Gods, Graves, and Scholars*), whose works brought the passion and excitement of the glory days of microbiology and archaeology, respectively, to many readers. The stories I tell were chosen for both their dramatic content and scientific importance. I have, I confess, "cherry-picked" the rich lore of natural history for the best of the best.

Ceram described adventure as "a mixture of spirit and deed." These stories are intended to capture those two elements in a compact form. They are crafted for enjoyment, not as scholarly biographies or histories of science. I have not delved into the biographical depth that would be necessary in full-length treatments. I have provided some background where I thought it would offer insights into who or what kindled the spirit of adventure in these naturalists and scientists.

Wherever possible I relied on field notes, journals, expedition reports, and other firsthand accounts because they tend to contain the person's thoughts and reactions at critical moments. I also examined original scientific papers because they are the material of record of what exactly was found, concluded, and proposed. Many of the individuals or discoveries described here have also been the subject of one, several, or many excellent books, some written in the first person and others by biographers. You will find the many sources I relied on in "Sources and Further Reading," at the back of the book, and I certainly encourage you to explore them. I had a blast researching these stories.

This book is not a compendium of the greatest evolutionary scientists or discoveries (although many here would certainly be on any such list), nor is it a history of the field. But the individuals here do represent very well the spirit of the enterprise, and many scientists, myself included, have drawn inspiration from one or more of them. So if you sense a lack of objectivity and a hagiographic bent, I am guilty as charged. There is much to admire in the protagonists of these stories. I have not paid attention to whether someone was a good citizen, good with money, or even a good spouse (some were, some weren't). These people had (or have) very rewarding and satisfying lives because of the great pleasure they gained from what they have done. I did not want to write a book about miserable bastards (although, come to think of it, that might be fun and would make for a catchy title).

The search for origins

The scientific quest driving all of these stories is the search for the origins of species, what was referred to by early scientists and philosophers as "the mystery of mysteries," "the question of questions," or the "supreme problem of biology." I will begin my tales and set the stage for the main body of the book with a short account of one of the boldest and most important scientific voyages ever undertaken. I am not referring to Darwin's famous journey but to the expedition through South and Central America made three decades earlier by Alexander von Humboldt (Chapter 1). It has been said that all scientists are descendants of Humboldt, who in the course of his travels made contributions to virtually every branch of science. We will see, however, that the magnificent flora, fauna, and fossils were perceived very differently by this great explorer than by those who followed him. Though brilliant, Humboldt and his illustrious circle of friends belonged to an age that embraced a religious vision of nature. Humboldt thus offers a view of the world before the revolution, and though he did not perceive a solution to the mystery of mysteries or even that the problem of origins could be solved, his journey blazed the trail for and directly inspired the wave of naturalists who did.

The main body of the book is organized into three parts, each of which focuses on one major aspect of the search for origins—of species in general (Part I), of particular kinds of animals (Part II), and of humans (Part III). Each part is preceded by a brief preamble that provides some background for the stories within it, and I have ordered the chapters in such a way that the connections between the scientists, discoveries, and ideas are highlighted in successive chapters. In Part I ("The Making of a Theory"), we follow the epic voyages of Charles Darwin and Alfred Wallace, who solved the origins problem, and of Henry Walter Bates, who discovered superb evidence of the process of natural selection. In Part II ("The Loveliest Bones"), we retrace several of the greatest expeditions and most spectacular finds in paleontology, which have thrown light on the origins of the animal kingdom and various major groups within it. And, in Part III ("The Natural History of Humans"), we track some discoveries from the archaeological and fossil records and explore new insights from the DNA record that have shaped our understanding of human origins.

This book debuts, not coincidentally, on the anniversaries of several milestones in natural history and evolutionary science. In 2009, we mark Charles Darwin's 200th birthday and the 150th anniversary of the publication of his *On the Origin of Species*. It is certainly fitting to celebrate the ideas and achievements of our greatest naturalist and the leader of this scientific revolution, and this book is in part my contribution to that party. We also mark the 100th anniversary of Charles Walcott's discovery of the remarkable animals of the Burgess Shale, which documented the Cambrian Explosion (Chapter 6), as well as the 50th anniversary of Mary and Louis Leakey's first ancient hominid find, which redirected the study of human origins back to Africa (Chapter 11).

But my aim goes well beyond these famous events. There is, for example, another, less well known anniversary to recognize, and its being so less well known is a matter of some concern that I hope to redress in my own small way. One hundred and fifty years ago, on July 1, 1858, the fruit of Darwin's *and* Wallace's great adventures— the theory of natural selection—was first presented publicly before a small audience of the Linnean Society in London and subsequently published in the society's journal. For reasons not altogether clear, nor for which there would be a scholarly consensus, this event and Wallace's contributions to it tend to be overlooked. Indeed, the most widely used college biology textbooks typically offer many pages about Darwin's travels and work and just a few vague lines about Wallace: "a young British naturalist working in the East Indies who had developed a theory of natural selection similar to Darwin's." Wallace, I think, is in danger of disappearing from textbooks altogether. And that, I hope you will agree after reading Chapter 3, would be a shame.

It is not a shame just because of the issue of credit in history. It is a shame because we are deprived of a great story of spirit and deed —Wallace's two long journeys, with a shipwreck in between, and his dozen years in the forests of the Amazon and Indonesia, and how and why he came to similar ideas as Darwin did while toiling on the other side of the globe. It is an inspiring tale of great passion, dedication, physical risk and endurance, perseverance, and the immense pleasure of discovery. There is so much to learn from his struggles and triumphs and to admire in his character.

The same can be said of all the naturalists and scientists in this book. In fact, one of the most important experiences shared by virtually every person in these stories is that their discoveries and ideas were initially rejected or doubted. One might have thought that finding the first ape-man or a new dinosaur, or deciphering some critical piece of history in DNA, would confer instant glory. Think again. Many struggled for decades before gaining widespread acceptance and recognition. Such is the nature of breakthroughs and revolutions in science.

What also unites the naturalists here unites all of us as humans— the urge to explore. New discoveries about the origins of our species reveal that most of us are the descendants of explorers, humans who migrated out of Africa just 60,000 years ago or so and eventually populated six continents (Chapter 13). Even if it may now be only from the safety of our armchairs or theater seats, we share a deep need to know the world around us.

In July 1976, on the eve of the historic landing of the *Viking I* spacecraft on Mars, NASA assembled a panel of luminaries, including the authors Roy Bradbury and James Michener, the physicist Philip Morrison, and the undersea explorer Jacques Cousteau, to discuss the motives for exploration. Most saw it as a matter of human instinct.

Cousteau offered, "What is the origin of the devouring curiosity that drives men to commit their lives, their health, their reputation, their fortunes, to conquer a bit of knowledge, to stretch our physical, emotional or intellectual territory? The more I spend time observing nature, the more I believe that man's motivation for exploration is but the sophistication of a universal instinctive drive deeply ingrained in all living creatures." Morrison agreed that "it is in our nature" and that "human beings explore because in the long run . . . both by genetics and by culture, we can do nothing else."

So the last thing I would want to imply through these stories is that the best days of scientific exploration are behind us. Far from it. Several of the stories here unfolded in just the last few years. Headlines are being made on a regular basis by newly found hominid, animal, and plant fossils, and many more surprises are still buried in the earth's crust. Powerful new tools for mining the DNA record of life and human evolution are certain to greatly expand our knowledge of our own natural history. There will be many new stories to tell.

But, you may ask, are we going to find anything that would turn our worldview upside down, anything of the magnitude of the revolution in thought that began 150 years ago? Is there anything still unknown that is comparable to "the mystery of mysteries" that launched these explorations to all corners of the world?

I think there is. And I will ponder that possibility in the Afterword.

Figure 1.1 Humboldt in Mexico. View of a rock formation and waterfall, with Humboldt and companions at lower left. From A. von Humboldt, *Vues des Cordillères et Monuments des Peuples Indigènes de l'Amérique* (1808).

∽ 1 ∽
Introduction: Humboldt's Gifts

Do not go where the path may lead;
go instead where there is no path and leave a trail.
— *Ralph Waldo Emerson*

RALPH WALDO EMERSON would label him "one of those wonders of
the world, like Aristotle . . . who appear from time to time, as if to
show us the possibilities of the human mind." Edgar Allan Poe would
dedicate his last major work, "Eureka," a 150-page prose poem, to
him. He so dazzled Thomas Jefferson that the two men began a lifelong correspondence. Though he never saw the American West, no
fewer than thirty-nine towns, counties, mountains, bays, and caves
were named for him, and the state of Nevada was nearly so. Only
Napoleon would be as famous in his time.

But that was all yet to come. First, the young Prussian naturalist
Alexander von Humboldt and his companion, the French botanist
Aimé Bonpland, would have to get their bearings in the South American paradise in which they had landed in the summer of 1799. No
naturalist had ever been there; all that lay before them was new and
unexplored. Humboldt wrote to his brother back in Europe:

"What trees! Coconut trees, fifty to sixty foot high . . . with enormous leaves and scented flowers, as big as the palm of a hand, of
which we knew nothing . . . And what colors in birds, fish, even cray-
fish (sky blue and yellow)! We rush around like the demented; in the
first three days we were quite unable to classify anything; we pick up

one object to throw it away for the next, Bonpland keeps telling me that he will go mad if the wonders do not cease soon."

The wonders did not cease, and thankfully Bonpland did not go mad, despite some of Humboldt's exploits.

Humboldt was curious, and well read, about *everything*. Two years earlier, while still in Europe, he had completed several thousand experiments that confirmed Luigi Galvani's discovery that muscle and nerve tissue are electrically excitable. His interest in animal electricity was thus very fresh when he encountered the wonder of "electric eels" in the streams in the Calabozo region of central Venezuela. Mesmerized by the three- to five-foot-long fish that he and his native assistants had brought to shore, he mistakenly stepped on one—and received an agonizing shock. "I do not remember ever having received a more dreadful shock from the discharge of a Leyden jar [an early device for conducting experiments in electricity], than which I experienced . . . I was affected for the rest of the day with a violent pain in the knees, and in almost every joint."

The eel's 500-volt wallop did not dissuade Humboldt from various experiments: "I often tried, both insulated and uninsulated, to touch the fish, without feeling the least shock. When M. Bonpland held it by the head, or by the middle of the body, while I held it by the tail, and, standing on the moist ground, did not take each other's hand, one of us received shocks while the other did not . . . If two persons touch the belly of the fish with their fingers, at an inch distance, and press simultaneously, sometimes one, sometimes the other will receive the shock."

He was game as well for many other potentially unpleasant experiments. When he and Bonpland discovered a "cow tree," a relative of the rubber tree named for its production of a kind of milk, Humboldt drank a gourd full of the liquid—much to Bonpland's horror. When a servant repeated the act, the poor fellow "vomited up rubber balls for several hours." Humboldt also tasted curare, the lethal poison used by the Indians to tip their blow-darts. Figuring it was deadly only if taken intravenously, Humboldt found "its taste of an agreeable bitter."

Such was the science of natural history in 1800.

But Humboldt's interests and aptitude spread far beyond these few curiosities. He was well versed in every domain of science—botany, geography, astronomy, geology—and in every aspect of human history in the New and Old World. In the course of his five-year journey (1799–1804) through Venezuela, Brazil, Guiana, Cuba, Colombia, Ecuador, Bolivia, Peru, and Mexico, he and Bonpland collected vast quantities of botanical, zoological, geological, and ethnographic specimens; made countless highly accurate maps; and witnessed a total eclipse, an earthquake, and a spectacular meteor shower (Leonid). They measured mountains and scaled the highest peak in Ecuador to an altitude of 19,286—higher than any human had ever gone before (even in a balloon) and a feat that would not be topped for 80 years; descended into a volcano; noted the cold north-flowing Pacific current that now bears Humboldt's name; and studied and admired the ancient civilizations of the pre-Columbian world (then unknown in Europe).

Such experiences were only made possible, of course, by Humboldt and Bonpland's exposing themselves to many dangers. Indeed, Humboldt was pessimistic about his odds of surviving the expedition, for he was convinced that his destiny was to be drowned on the high seas. After five years during which he escaped attacks from natives, dodged ambush by jaguars, withstood the endless assaults of swarms of mosquitoes, battled tropical diseases, endured jailing by authorities, and somehow avoided drowning when his canoe capsized (he could not swim), Humboldt's destiny was almost realized on his return journey.

When Humboldt stopped in Cuba, an American diplomat encouraged him to delay his return to Europe and to visit the United States. Humboldt was already an admirer of Thomas Jefferson and decided to make the detour. But in May 1804, on his way to Philadelphia from Havana, his ship was caught in a bad storm off the coast of Georgia. Humboldt feared for his life. After all he and Bonpland had endured, he despaired that they might die so close to the conclusion of their voyage. He later wrote in his journal:

I felt very much stirred up. To see myself perish on the eve of so many joys, to watch all the fruits of my labors going to pieces, to cause the death of my two companions [a young Ecuadoran accompanied Bonpland and Humboldt on their trip], to perish during a voyage to Philadelphia which seemed by no means necessary . . .

Once the storm abated, the ship had to pass through the British naval blockade that then extended across all of the harbors along the east coast.

Finally safe in Philadelphia, Humboldt took to the new republic very quickly. He believed that America was a great new country that was freeing itself from the shackles of the outmoded European order. He wrote directly to Jefferson both to introduce himself and to state his purpose in visiting. After a warm, flattering greeting and a brief description of his past five years of travels, he told the president: "I would love to talk to you about a subject that you have treated so ingeniously in your work on Virginia, the teeth of a mammoth which we discovered in the Andes of the southern hemisphere at 1,700 toises [about 10,800 feet] above the level of the Pacific Ocean."

Yes, that's right, Humboldt wanted to talk to the principal author of the Declaration of Independence, the former governor of Virginia (1779–1781), the first secretary of state (1789–1793), the second vice president of the United States (1797–1801), and its third president— about fossils.

Humboldt knew of Jefferson's keen interest in fossils, particularly those of the mammoth, from his *Notes on the State of Virginia* (1785). The work began as a response to a list of questions from the French concerning the fledgling states they were assisting. It grew to be a comprehensive treatment of the geography, flora, fauna, agriculture, history, customs, commerce, and other subjects concerning Virginia and the United States.

In *Notes*, Jefferson had written about the "mammoth" bones that had been discovered in Kentucky and New York's Hudson Valley. He used their existence to refute the so-called Theory of American Degeneracy propounded by a French naturalist, the Comte de Buffon. Buffon had alleged that the more humid and colder climate of North America, compared to that of Europe, led to a marked inferiority in its wildlife, livestock, and indigenous people. That did not sit well at all with Jefferson, who emphasized the great size of the mammoth— "the largest of all terrestrial beings"—and thought it alone was sufficient to squash Buffon's theory.

Years later, while vice president, Jefferson was given a set of bones from a cave in West Virginia. He analyzed a forelimb and a hand with giant claws and dubbed the unknown animal "Megalonyx," or "Giant-Claw." He first thought that they might belong to a giant cat three

Figure 1.2 **Chimborazo volcano.** Humboldt ascended the peak at left to the 19,286-foot level, the highest elevation attained at the time. From A. von Humboldt, *Vues des Cordillères et Monuments des Peuples Indigènes de l'Amérique* (1808).

times the size of a lion. He was not confident of his interpretation, however, and when he caught a glimpse of a giant ground sloth in an article by the French paleontologist Georges Cuvier, he saw a possible resemblance. His bones were in fact those of a ground sloth, which was later named *Megalonyx jeffersonii* in his honor.

In 1799, Jefferson's article on Megalonyx was published in the *Transactions of the American Philosophical Society,* perhaps the first American publication in the field of paleontology. But far more important to Jefferson than scientific credit was what the mammoth and the Megalonyx bones signified about his expanding republic. At the time, the idea that fossils represented extinct species was not well established. Jefferson and many others could not accept that any link in God's chain of creation would be allowed to perish. "Such is the economy of nature," Jefferson wrote, "that no instance can be produced of her having permitted any race of animals to become extinct." He believed that mammoths and other beasts were still roaming:

In the present interior of our continent there is surely space and range enough for elephants and lions . . . Our entire ignorance of the immense country to the West and North-West, and of its contents, does not authorise us to say what it does not contain.

A few years later, as president, when he sent Lewis and Clark west, he included the order to observe "all the animals of the country generally, & especially . . . any which are deemed rare or extinct." After their return, he then personally financed an expedition led by Clark to a large deposit in Kentucky, which yielded several hundred mammal bones. About half of them were sent to the White House, where they filled the unfinished East Room, dubbed at the time the "Bone" or "Mastodon" Room.

When Humboldt arrived in Washington in June 1804, he met with Jefferson, Vice President James Madison, and other officials over a span of ten days. There was talk of fossils, but Jefferson had many other concerns on his mind. The recently completed Louisiana Purchase now gave the United States a border with Spanish America. Jefferson was desperate for intelligence on Mexico when Humboldt appeared—fresh from Mexico with accurate data and maps and insights into its political and economic situation. Humboldt had all of the answers to Jefferson's questions about roads, mines, Indian tribes,

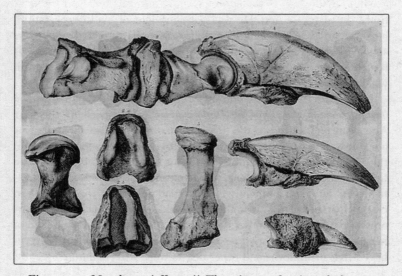

Figure 1.3 *Megalonyx jeffersonii.* These bones of a giant sloth were found in Virginia and described by Thomas Jefferson in *Transactions of the American Philosophical Society,* vol. 4 (1799). The drawings are by James Akin. The image is from the Rare Book and Special Collections Division of the Library of Congress.

crops, settlements, and more, and he was delighted to share it with perhaps the only man whose reach and grasp matched his own.

Humboldt thereafter showed his admiration for Jefferson, and all things American, in many ways. After he returned to Europe, he received a steady stream of American dignitaries—diplomats, politicians, inventors, and writers—over the next few decades. Most important, Jefferson's *Notes on the State of Virginia* served as a model for Humboldt's writings. The American's full descriptions of his country, its geography, people, and history, its animals and plants, its climate and commerce, influenced Humboldt's approach to documenting his travels in the New World.

Those writings would encompass every dimension of the countries Humboldt visited, as well as the heavens above them. The full account of his voyage comprised some thirty volumes, which were published over the ensuing three decades. The 1,425 maps and illustrations in his opus were so elaborate that the cost of reproducing them would eventually bankrupt him. In his later years, Humboldt maintained his prodigious output. At the age of seventy-six, he published the first part of his five-volume *Kosmos: A Sketch of the Physical Description of the Universe* (1845), "an attempt to delineate nature in all its vivid animation and exalted grandeur." His books and frequent contact with the leaders and most prominent citizens of the New and Old Worlds made him famous the world over.

Humboldt was, as one historian described, "the master of all branches of science at the last moment in history when this was possible for a single human being." After Humboldt, even the greatest naturalists would be specialists.

And because of Humboldt, there came many naturalists.

The shorter account of his voyage, his *Personal Narrative of Travels to the Equinoctial Regions of the New Continent* (1815), not only enlarged his fame but also inspired many of the major figures of nineteenth-century natural history and exploration—names that are, ironically, better known today than Humboldt's. Young Charles Lyell, who became the father of modern geology, remarked after meeting the great Humboldt in Paris, "There are few heroes who lose so little by being approached as Humboldt." Humboldt also directly supported the young Swiss naturalist Louis Agassiz and convinced him to move to the United States, where he became a professor at Harvard and the leading figure in American natural history.

Then there was Charles Darwin. As a fledgling student at Cambridge University in the late 1820s, Darwin read all seven volumes and 3,754 pages of Humboldt's *Personal Narrative.* He was so enamored of Humboldt's descriptions of the Tropics that he read them over and over, committing parts to memory and reciting them aloud until he drove his friends crazy. The first volume of the *Personal Narrative* was one of the few books that Darwin took with him on the voyage of the *Beagle,* and he read it often to buck up his courage to endure his constant seasickness. It became the model for his own travelogue, *The Voyage of the Beagle.*

Humboldt's tales of his South American adventures had the same effect on Alfred Wallace and Henry Walter Bates, inspiring their decision to explore the Amazon region.

But, as great as Humboldt's accomplishments and influence were, his particular view of nature was to be overturned by the very generation of naturalists he inspired. Humboldt saw nature, both its living and nonliving components, as a somewhat static and peaceful domain that reflected an integrated design and divine order. Before his expedition, he wrote to a colleague that his main purpose was "to observe the interactions of forces, the influence of the inanimate environment on plant and animal life. My eyes will constantly focus on this harmony."

Humboldt did not seek explanations of, for example, the origins of living things, for he saw that question as outside the sphere of natural history. So, too, did many of his contemporaries. Agassiz defined a species as a "thought of God" and declared that "Natural History must in good time become the analysis of the thoughts of the Creator of the Universe."

Those who followed Humboldt into the Tropics developed an altogether different agenda for natural history and a picture of nature as a perpetual struggle among all organisms, a view that would replace entirely the Humboldtian ideal.

Humboldt died in May 1859, just six months before this new world view was articulated in *The Origin of Species.*

The eminent historian David McCullough has suggested that Humboldt's most important impact was to demonstrate "how relatively little had been known of the richness and variety of life on Earth, the infinite abundance of life's forms, and how infinitely much more there was to know." To young bug collectors who had exhausted the limited resources of damp, cold, gray England, the destinations of Humboldt's *Narratives* were irresistible. Add the high adventure and romance of traveling in unexplored regions and the thrill of encountering natural wonders, and it is perhaps no wonder why so many followed in his footsteps.

PART ONE

THE MAKING OF
A THEORY

Rule Brittania! Brittania, rule the waves!

At no time was the rousing chorus of this centuries-old British patriotic song more true than in the first half of the 1800s. As other nations recovered from war and political upheavals, Britain's navy was able to command the seas and to help expand trade throughout its far-flung, thriving empire. A system of armed "packet ships" and, later, commercial vessels carried mail and parcels to, from, and between many outposts around the globe. If one had the inclination to cross the seas to explore faraway places and collect exotic specimens, it helped to be British.

Three great voyages and three great British naturalists were pivotal to the conception of the idea of natural selection and the development of evidence to support it. The best known voyage of all is, of course, Charles Darwin's on the HMS Beagle. *While Darwin is justifiably famous for his contributions to natural history and the theory of evolution, the turn of events that put him on that ship, his outlook and motivations, and how he came to see the world so differently are not as widely known and are often misconstrued. The insecure divinity student who boarded the* Beagle *was a most unlikely future revolutionary.*

Darwin did not set out with any intention to gather evidence for or against any great idea. His theory of evolution took shape after the voyage as he pondered what he had seen and begin to question (privately) the thinking at the time. Alfred Russel Wallace and Henry Walter Bates, on the other hand, did have evolution in mind from the start. The idea that species might change was percolating in popular circles in the mid-1840s. It was Wallace who suggested to his friend Bates that they go to the Amazon to gather data "toward solving the problem of the origin of species."

All three were young men when they set out from England for the jungles of South America. Darwin was twenty-two, Bates, twenty-three, and Wallace, twenty-five. Unlike Darwin, however, who came from a wealthy family, was educated at Cambridge, and enjoyed the advantages of being the naturalist on an armed navy vessel, Bates and Wallace were self-taught amateurs who had to make their way to the Amazon on a commercial trading ship and then, to cover their expenses, shipped prized specimens back to England for sale. Once in the Amazon, the two friends soon split up to cover more territory. Wallace headed home after four years, then undertook a long solo voyage throughout the many islands of the Malay Archipelago. Bates stayed in the Amazon jungle for eleven difficult but rewarding years.

These were truly epic journeys filled with harrowing and joyous moments. Prodigious collectors of bugs, birds, and anything else that moved, all three men developed a great appreciation for the diversity of species, for variation within species, and for the geographic distribution of species and varieties. It was primarily this appreciation that led each of them to their respective discoveries: Darwin to his idea of "natural selection" and the descent of species from common ancestors (Chapter 2); Wallace to his independent conception of the "struggle for existence" among individuals and the "Wallace line," demarcating the boundaries between Asian and Australian animals (Chapter 3); and Bates to his discovery of the phenomenon of mimicry, which provided the best and very timely evidence for natural selection in the wild (Chapter 4). The theory of evolution would, after The Origin of Species, *be forever linked to Darwin, but the making of the theory and its early scientific acceptance owe a considerable debt to Wallace and Bates, respectively.*

Although Darwin's journey (1831–1836) preceded those of Wallace (1848–1862) and Bates (1848–1859) by almost twenty years, their work intertwined when the latter two men returned to England. The three explorers were friends and correspondents for the rest of their lives.

Figure 2.1 "I was in many ways a naughty boy." Portrait of young Charles and his sister Catherine. Charles later wrote in his autobiography: "I was much slower in learning than my younger sister Catherine, and I believe that I was in many ways a naughty boy." From *More Letters of Charles Darwin: A Record of His Work in a Series of Hitherto Unpublished Letters*, edited by F. Darwin and A. Seward (D. Appleton and Co., New York, 1903).

∿ 2 ∿
Reverend Darwin's Detour

Every traveller must remember the glowing sense of happiness,
from the simple consciousness of breathing in a foreign clime,
where the civilized man has seldom or never trod.
— *Charles Darwin*, Voyage of the Beagle *(1839)*

His nickname was "Gas."

Thirteen-year-old Charles Darwin was, like most younger brothers, highly susceptible to conspiracy and mischief with his older brother. Erasmus ("Ras"), five years his senior, had developed an interest in chemistry and recruited his little brother into outfitting a makeshift lab in the garden shed. The two boys pored over chemistry manuals and often stayed up late, concocting some noxious or explosive mixture.

Sons of a wealthy doctor, Ras and Gas always had plenty of funds for their hobbies. They bought test tubes, crucibles, dishes, and all sorts of other apparatus. Of course, chemistry wasn't as much fun without fire, so the boys invested in an Argand lamp, a type of oil lamp that they used to heat chemicals and gases. Their fledgling laboratory also had fireproof china dishes, courtesy of their uncle Josiah Wedgwood II, the leading maker of pottery in England.

Charles enjoyed the prestige his stinky shed earned him among his schoolmates. He was also popular for his cheerful and mild-mannered disposition. Some joined him on his expeditions into the countryside collecting insects or bird-hunting. The boarding school he attended was only a mile from home, so he knew the surrounding woods and streams very well.

The schoolmaster, however, was not so impressed with either Charles' chemistry or his lackadaisical approach to the classics. Charles was not much of a student. He was bored stiff by the rote learning of ancient geography, history, and poetry demanded by the school. He escaped to the woods and home to visit and play with his dog as often as he could, sometimes risking expulsion should he get locked out before bedtime. Racing back to his dormitory, he prayed out loud for God's help in getting him there in the nick of time, and he marveled that his prayers were answered.

His father was increasingly aware of Charles' dislike for school. Although Charles adored him, Robert Darwin was a large, imposing figure and what he said governed the Darwin household. "The Doctor," as he was called, was worried that Charles was frittering away his opportunities. One day, his anger erupted: "You care for nothing but shooting, dogs, and rat-catching, and you will be a disgrace to yourself and your family!"

The Doctor decided that the best thing would be to take Charles out of school two years early, at the age of sixteen, and send him to Edinburgh, where he could stay with Ras and enroll in medical school. The Doctor hoped that Charles would follow in his footsteps, and those of his grandfather, and become a physician.

Surgery and sponges

In Edinburgh, Charles learned many things—taxidermy, natural history, zoology, and that he did not want to be a doctor.

The university's medical school provided the best training in Great Britain, but it was a gruesome ordeal in the 1820s. Charles was repulsed by the professor of anatomy, who showed up for his lectures dirty and bloody, fresh from stints at the dissecting table. Charles also found surgery sickening. In these days before anesthesia, speed was essential, and the procedures did not look too different from those in a butcher's shop. After witnessing an operation on a child, Charles fled the operating theater and vowed not to return.

Disgusted by some classes, bored by others, Charles started to find other diversions instead of attending lectures. When the Doctor caught wind of Charles' waning interest, he sent a message via Charles' sister Susan:

He desires me to say that he thinks your plan of picking & chusing what lectures you attend, not at all a good one . . . it is quite necessary for you to bear with a good deal of stupid & dry work; but if you do not discontinue your present indulgent way, your course of study will be utterly useless.

Outside the horrors of medical school, Edinburgh did offer attractive excursions. Charles loved to walk along the dramatic coastline of the Firth of Forth and to look for any sea creatures that washed ashore. In the city, he met a freed slave from Guiana, John Edmonstone, who agreed to tutor him on how to stuff and mount birds. Charles was an excellent student and reveled in Edmonstone's tales of the Tropics, and his descriptions of the South American rainforests were perfect antidotes to the bone-chilling Scottish climate.

In the summer after his first year, Charles was relieved and happy to return home and to stalk the nearby woods once again. He made some stabs at continuing his medical education. His father encouraged him to read his grandfather Erasmus Darwin's book on life and health, entitled *Zoonomia; or, the Laws of Organic Life*. In this multi-volume tome, Grandfather Erasmus had opined on topics ranging from the basis of disease to the history of life. On the latter subject, he was unconventional, to say the least:

If this gradual production of species and genera of animals be assented to, a contrary circumstance may be supposed to have occurred, namely, that some kinds by the great changes of the elements may have been destroyed. This idea is shewn to our senses by contemplating the petrifactions of shells, and of vegetables [plants], which may be said, like busts and medals, to record the history of remote times.

While Charles no doubt admired his grandfather's book, the greater philosophical messages were probably lost on the youth.

During his second year in Edinburgh, he drifted still further from medicine and much closer to natural history. But he did find one professor much to his liking, the zoologist Robert Grant, who was an expert on—some might say quite mad about—the marine animals that abounded in the tide pools near Edinburgh. Grant's boundless enthusiasm and sense of humor won Charles over, and they became frequent walking companions. Grant taught Charles what to look for, and Charles diligently took notes on Scottish sponges, molluscs, polyps, and sea pens.

Grant was well traveled, well read, and a free thinker. He rejected the prevailing orthodox view in Britain of the fossil record's documenting a series of episodes of creation and each species as specially created and unchanging. Grant was a follower of the French naturalists, who thought that life did change as a product of natural laws. He introduced Charles to the work of Jean-Baptiste Lamarck and his ideas about the inheritance of acquired characteristics. He also took Charles to meetings where these topics were fiercely debated.

Grant showed Charles how to ask both small and great questions and the connection between the two. But Charles was no closer to being a doctor—or anything else, for that matter. He dropped out of medical school without a degree.

The making of a country parson

The Doctor had to find something respectable for his aimless son. There was a great fear among the well-to-do that their privileged sons would be content to live off the family fortune. If not medicine or the law, what would befit Charles? What position would bring him the most respectability for the least ambition? The Church of England.

It was then a common practice for parishes to be auctioned off to the highest bidder, who would then install a family member as parson. It was a comfortable way of life with ample lodging and some land, along with income from the parishioners and investments. Charles would have plenty of time to pursue his hobbies.

He would need only to pass his ordination, which required a bachelor's degree from Cambridge or Oxford, and a year's study of divinity. So it was off to Cambridge for Charles, where virtually all of the faculty were ordained members of the clergy.

Having washed out of Edinburgh, Charles was determined to make a good start. Unfortunately, his resolve was soon challenged by the craze that was then sweeping the nation and Cambridge—beetle collecting. The capture of diverse and rare species was becoming a competitive sport. It so appealed to Charles' love for romping about the woods with like-minded comrades and his thirst for recognition that he soon became obsessed.

Charles acquired the best equipment, hired helpers to sift through forest debris, and spent significant sums buying specimens from other collectors. One day, peeling bark off a tree, he eyed two rare forms and quickly grabbed one in each hand, only to spot another. He popped one in his mouth so he could snag the third. Unfortunately, the one in his mouth was a bombardier beetle; it emitted an awful concoction that forced him to spit it out and lose the other two.

Charles spent most of his first two years pursuing such trophies but finally resolved again to work hard and prepare for the critical exam at year's end. He was to be tested on Latin and Greek translation, portions of the Gospels, the New Testament, and the works of the Reverend William Paley, who had written several books concerning the evidence for God and the truths of Christianity. Charles, in fact, was lodging in the very rooms Paley had occupied at Cambridge and was very impressed and persuaded by Paley's lucid logic.

Charles made it through the exams and resumed his beetling, but he also fell under the very positive influence of his professor of botany, the Reverend John Stevens Henslow. On Friday nights, Henslow had small gatherings at his home for the discussion of natural history and a little wine-drinking. Other professors would occasionally drop in and share their expertise and passions. Charles had found his home. Henslow took him under his wing, and the two were so often seen walking together, engrossed in conversation, that Gas became instead "the man who walks with Henslow."

Henslow took students on all sorts of botanical excursions around Cambridge. Charles was eager to please, even wading through the muck of the River Cam to snatch a rare species for his mentor. He saw Henslow as the role model of the ordained naturalist and admired him as "quite the most perfect man I ever met with."

Henslow was not the bold freethinker that Grant was. One of the requirements at Cambridge was adherence to the Thirty-Nine Articles of the Church of England (established in 1563), and Henslow supported every word of them. Charles thought he would take his year of divinity studies under his tutelage.

First, there was the matter of passing his final exam—more Homer, Virgil, and Paley—with some math and physics tossed in. Charles ranked tenth in a group of 178.

He had his degree, once he signed the Thirty-Nine Articles, and Henslow continued his grooming. He encouraged Charles to read more and to think about traveling to widen his horizons.

To get him started, he lent Charles his copy of Humboldt's *Personal Narrative*. Whereas the old Charles would have struggled with the seven-volume work, the new Charles gobbled it up and started dreaming of the places Humboldt described in his travels throughout South and Central America. The Canary Islands offered the closest tropical paradise, so Charles started plotting a trip. Henslow and three friends were initially interested in joining him. Charles' father put up the money to clear all his debts and to pay for the expedition.

Henslow knew that Charles would need some geological training to make the most of such a visit, so he set up a tutorial with the Reverend Adam Sedgwick, a professor of geology, who had earlier trained Henslow. Sedgwick, a leading figure in British geology who later named the Devonian and Cambrian periods, took Charles to Wales on a field trip. Charles discovered that he had a knack for and loved geology.

During this excursion, his Canary Island companions, including Henslow, backed out until just one remained. On his way home, Charles received a message that this last of his companions had died suddenly. He was shocked at the loss and deeply disappointed that his expedition had unraveled.

By the time he reached home, exhausted and uncertain of his next step, awaiting him was a letter from Henslow with stunning news. Charles was being offered a voyage around the world.

Permission to board

The invitation had come via a circuitous route. Captain Robert FitzRoy had received orders from the Admiralty to lead the HMS *Beagle* on a detailed surveying voyage around the southern portion of South America. He had taken the helm of the ship on its previous expedition when the captain committed suicide under the stresses of command. Well aware of the pressures on and the isolation of naval captains on such long expeditions, FitzRoy requested that a "well-educated and scientific person" join the voyage so that "no opportunity of collecting useful information . . . should be lost." It was not uncommon for naturalists to travel aboard naval ships, but FitzRoy was at least as interested in having a well-heeled companion for dinner and conversation as in what might be collected.

The Admiralty first offered the berth to Henslow and another naturalist. They declined, and each recommended Charles. Henslow wrote to his protégé: "I think you are the very man they are in search of."

Charles was elated. The Doctor was not.

Charles' father knew that British ships were sailed by some very rough characters. He also knew that they were dangerous and often wound up being sailors' coffins. He thought the voyage too risky an adventure and yet another delay on Charles' path to settling down in a respectable position. Charles glumly wrote to Henslow that he would not defy his father's objections.

Charles headed off to his uncle Josiah's house to distract himself from his disappointment. His father, however, handed him a letter for his uncle in which he explained that he objected to the voyage on many grounds, but "if you think differently from me I shall wish him to follow your advice."

The Wedgwoods were indeed supportive of the adventure. Uncle Jos asked Charles to write down a list of his father's objections so that he could offer his response. The words still fresh in his mind, Charles recalled:

1. Disreputable to my character as a Clergyman hereafter
2. A wild scheme
3. That they must have offered to many others before me, the place of Naturalist
4. And from its not being accepted there must be some serious objection to the vessel or expedition
5. That I should never settle down to a steady life hereafter
6. That my accommodations would be most uncomfortable
7. That you should consider it as again changing my profession
8. That it would be a useless undertaking

Uncle Jos wrote to the Doctor, who quickly changed his mind and said he would now give the voyage "all the assistance in my power."

Overjoyed and with little time to prepare, Charles set about buying instruments and a new pistol and rifle, getting packed, and meeting with Captain FitzRoy. The ship itself was a bit of a shock. The

HMS *Beagle* was *small*—only 90 feet long and 24 feet wide at the most—with just two tiny cabins (Figure 2.2). Charles, at six feet tall, had to stoop to enter what would be his quarters for years, which he would share with a large chart table, a nineteen-year-old officer, and a fourteen-year-old midshipman, Philip King. Charles was to sleep in a hammock slung over the chart table just two feet below a skylight.

Charles made his rounds, saying good-bye to friends and family and seeking out last-minute advice from naturalists. Henslow gave him a parting gift—a copy of Humboldt's *Personal Narrative*—and suggested that he take as well Lyell's new *Principles of Geology* to help him decipher the landscapes he would see. These books, and a copy of the Bible, would be the young divinity student's close companions.

Saying good-bye to his father was the hardest. He would be away for so long. The voyage was intended to last two years, but little did Charles or his father know that it would take five. There was also the very real chance he might not return. Charles tried to put these thoughts out of his mind as Ras saw him off from Plymouth. On December 10, 1831, the *Beagle* made what turned out to be the first of

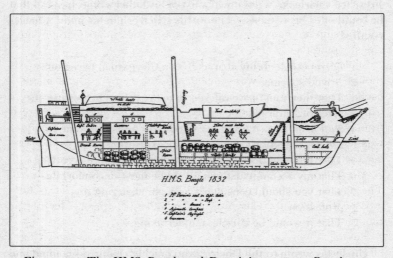

H.M.S. Beagle 1832

Figure 2.2 The HMS *Beagle* and Darwin's quarters. Based on a drawing by his shipmate Philip King, with whom he shared his quarters. From *Journal of Researches into the Geology and Natural History of the Various Countries Visited by H.M.S. Beagle*, by Charles Darwin (facsimile edition of 1839 first edition, Hafner Publishing Company, New York, 1952).

three attempts to start its voyage; it hit a strong gale and was thwarted, forcing FitzRoy to turn back. He then set out again at low tide on December 21, only to run the ship aground. Once freed, another gale turned them back. Charles spent most of Christmas Day on the ship. Finally, on December 27, 1831, twenty-two-year-old Charles Darwin and his shipmates set out for the Canary Islands and South America.

Voyaging

It did not take long for him to be miserable. As the *Beagle* tossed in the notorious waves of the Bay of Biscay, Charles tossed everything he tried to eat. He retreated to his hammock and wondered if the voyage was a huge mistake. He pulled out Humboldt for a little encouragement and tried to look forward to setting foot on land again.

After ten days of torment, the ship reached Tenerife, in the Canary Islands. Charles finally saw the great mountain of which Humboldt wrote, but his excitement was short-lived. The *Beagle* was to be quarantined for fear of its sailors spreading the cholera that had erupted in England. FitzRoy wasn't going to wait: he ordered the sails up, and the ship left for St. Jago in the Cape Verde Islands without anyone stepping onshore.

St. Jago provided relief from Charles' desperate seasickness. Although the landscape was volcanic, the birds, palms, and massive baobab trees deeply impressed Charles. He was also intrigued by a band of shells and corals that lay about 30 feet above sea level. Fresh from Sedgwick's training and his reading of Lyell, Charles began to wonder: Had the sea level fallen or the island risen? He would ponder the same questions many more times in the coming years.

After a few weeks, it was back onboard for the crossing to Brazil. The nausea returned, compounded by the oppressive heat as they crossed the equator. Charles was laid up in his cabin, feeling as though he was being "stewed in . . . warm melted butter."

Having retched his way across the Atlantic, he was eager to get off the boat as soon as it made landfall at Bahia, on the coast of Brazil. Charles headed for the forest, and it did not disappoint. His senses were flooded—the colors of the flowers, fruits, and insects, the scents of the plants and trees, and the chorus of all the animal sounds. He wrote to Henslow: "I formerly admired Humboldt, I now almost adore him; he alone gives any notion of the feelings that are raised in the mind upon entering the Tropics." Charles began to collect everything he could.

After several weeks in Bahia, the *Beagle* sailed on to Rio de Janeiro, and Charles again ventured out. This was to be the pattern of the voyage. The *Beagle* would sail from port to port, conducting its surveys and mapmaking, while Charles would head inland to collect. Captain FitzRoy was obsessive about his own work, which bought Charles a lot of time for his excursions.

Onboard ship, there was also a routine. In a letter to his sister, Charles explained:

> We breakfast at eight o'clock. The invariable maxim is to throw away all politeness—that is, never to wait for each other, and bolt off the minute one has done eating, &c. At sea, when the weather is calm, I work at marine animals, with which the whole ocean abounds. If there is any sea up I am either sick or contrive to read some voyage or travels. At one we dine. You shore-going people are lamentably mistaken about the matter of living on board. We have never yet (nor shall we) dined off salt meat . . . At five we have tea.

Charles himself procured a good portion of the meat his shipmates ate in the course of the voyage. He was, thanks to his boyhood, a good shot, and his skills put him in good stead with the crew of the *Beagle*.

Charles also used the stops to find a vessel heading home that could carry his specimens. Eight months into the voyage, he shipped his first box, to Henslow, for safekeeping.

Gauchos and bones

Forays into the interior required some local knowledge, and Charles was usually able to find various characters willing to accompany him. At Bahia Blanca, a settlement on the coast of Argentina at the edge of the great Patagonian plains, or Pampas, he found himself in the company of gauchos, the local cowboy that Charles found "by far the most savage picturesque group I ever beheld." Amused by their colorful dress and ponchos, Charles also noted the sabers and muskets they carried. They were in constant conflict with local tribes, but as they were also "well known as perfect riders" and knew where to find the few sources of fresh water, Charles and the officers of the *Beagle* accepted their assistance. This included an introduction to the local cuisine of rhea eggs (from a flightless bird Charles called an "Ostrich") and armadillos, which Charles declared "taste & look like a duck."

While exploring the coast just a bit farther south, near Punta Alta, Charles found some rocks containing shells and the bones of large animals. He used his pickax to free what he guessed were parts of a "rhinoceros." The next day he found a large head embedded in soft rock and spent many hours removing it. Two weeks later, he found a jawbone and a tooth that he thought belonged to the *Megatherium*, or giant ground sloth. He was not sure of what he had, so he crated the bones ("cargoes of apparent rubbish," FitzRoy teased) for shipment so that the experts back in England could decipher their identities.

Eventually, it was determined that Charles had found remnants of several species, including a giant armadillo-like creature called a glyptodont, *Toxodon*, an extinct relative of the capybara, and three types of ground sloth — *Megatherium*, *Mylodon* (Figure 2.3), and *Glossotherium*.

It would be a long wait before Charles would hear that his fossils had arrived safely. The safe passage of either parcels or passengers was far from guaranteed in those days, as Charles was about to learn firsthand.

Land of savages

The *Beagle* continued to sail south along the eastern coast of South America, making its way toward Tierra del Fuego and Charles' first encounter with humans in their most primitive state.

He had been looking forward to the experience. On the previous voyage, Captain FitzRoy had taken several Fuegians back to England, where they were clothed and taught in the British fashion. Now, three of these former "savages" were to be returned to their people in the hope that they might spread some civilization to this part of the world.

Charles rowed ashore with FitzRoy to meet the natives. He was shocked by their appearance and behavior and could not believe that the three missionaries they were about to deliver were only recently just as untamed (Figure 2.4). The contrast set in motion much

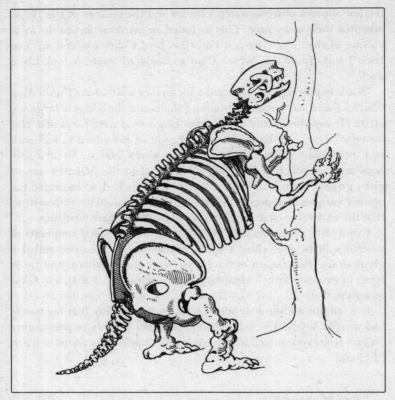

Figure 2.3 A mylodon. Drawing of a giant ground sloth from *A Naturalist's Voyage Around the World: The Voyage of the H.M.S. Beagle*, by Charles Darwin (D. Appleton and Co., New York, 1890).

thought about the differences, or lack thereof, between savage and civilized humans.

The *Beagle* forged on to round the notorious Cape Horn. Hugging the coast, the ship often tucked into coves to escape the weather. Charles tried to enjoy the scenery and wildlife, but two weeks of the cold, wind, and waves took their toll. He noted in his diary: "I have scarcely for an hour been quite free from seasickness: How long the bad weather may last, I know not; but my spirits, temper, and stomach, I am well assured, will not hold out much longer."

But when the weather worsened, the *Beagle* lost track of its position and took a pounding. A great wave struck the ship, and the crew had to cut away one of the whale-boats. The sea poured onto the decks and started to fill the cabins. Fortunately, once the portholes were opened, the little ship righted herself and the water drained away. One more wave, Charles knew, would have been the end. It was the worst gale FitzRoy had ever experienced. Terrified, Charles wrote in his diary: "May Providence keep the *Beagle* out of them."

They inched up the coast to establish a settlement for their Fuegian missionaries. Afterward, they entered the Beagle Channel. The scenery was magnificent. Glaciers extended from the mountains down to the water, where they calved small icebergs. But the tranquility of the scene was deceiving. While a landing party dined onshore near a glacier, a large ice mass broke off and hit the water, sending a great wave toward their boats on the shore. Charles and several sailors acted quickly, grabbing the boat lines before the waves could steal them away. Had they lost the boats, they would have been in a dire situation, stranded with no supplies in hostile country.

FitzRoy was impressed by Charles' actions and the next day named a large body of water "Darwin's Sound" after "my messmate, who so willingly encountered the discomfort and risk of a long cruise in a small loaded boat." The captain also named a mountain peak in Charles' honor.

Charles certainly appreciated these gestures. It was flattering for the twenty-four-year-old geologist to have features named for him, even if they were at the remote tip of the South American continent.

But, as the second year of the voyage unfolded and Charles continued his expeditions and collecting, he was increasingly concerned with how his efforts were being received back in England. He had shipped home more fossils, including a nearly complete *Megatherium*, and barrels of specimens. Given the long wait between sending a shipment or letter and receiving a reply, Charles was worried. Had his shipments even reached Henslow? Was he collecting anything of interest? His constant seasickness and bouts of homesickness were also wearing on him. He confessed in a letter to Henslow his anxiety about the length of the voyage: "I know not, how I shall be able to endure it."

When the *Beagle* arrived in the Falkland Islands in March 1834, mail was waiting, and Charles finally got his answer. In a letter composed six months earlier, on August 31, 1833, Henslow reported that

Charles' fossil *Megatherium* "turned out to be most interesting" and had been shown at the meeting of the British Association for the Advancement of Science that summer. The mentor gently encouraged his pupil: "If you propose returning before the whole period of the voyage expires, don't make up your mind in a hurry . . . I suspect you will always find something to keep up your courage." Then he added: "Send home every scrap of *Megatherium* skull you can set your eyes upon — *all fossils* . . . I foresee that your minute insects will nearly all turn out new."

Henslow's news and encouragement were just what Charles needed. He returned to his geology and collecting with zeal and looked forward to the next sights on the voyage — the west coast of South America and the Andes.

Shaky ground

The price for every new adventure was yet another confrontation with the sea. To get to the west coast, the *Beagle* sailed through the Strait of Magellan — a "short-cut" that avoided the treacherous Cape Horn (Figure 2.5). But, in late May and early June, it was no leisurely cruise. Charles watched the ice form on his skylight while he clung to his hammock.

There were many reminders of the peril of each leg of the voyage. On the way north to Chile, a shipmate died and was buried in a solemn service at sea. Later, as the *Beagle* scouted the islands off the Chilean coast, they glimpsed a man waving a shirt, and a party was sent ashore to investigate. They found five American crew members who had run away from a whaling ship in a small boat and wrecked before they could reach the mainland. Charles saw that the men were in desperate shape, having survived for over a year on nothing but shellfish and seal meat.

On the mainland, Charles enjoyed more geological excursions. He found beds of modern shells at an elevation of 1,300 feet, and in the Andes he found fossil shells at 13,000 feet. How could the marine creatures have wound up so high above the sea?

In a forest near Valdivia, he got part of his answer. While resting on a morning walk, he felt the earth tremble, then shake so violently he could not stand. He went back into town to find chaos. Houses were tilting, and the citizens were in shock.

Figure 2.4 A Fuegian. Drawing at Portrait Cove from *Journal of Researches into the Geology and Natural History of the Various Countries Visited by H.M.S. Beagle,* by Charles Darwin (facsimile edition of 1839 first edition, Hafner Publishing Company, New York, 1952).

As the *Beagle* sailed north, they saw devastation everywhere. The city of Concepcion had been pummeled to rubble. The inhabitants described the earthquake as the worst ever; it had also triggered a tsunami and widespread fires. Many people were still buried.

At the shore, Charles observed that the mussel beds were now positioned several feet above the water. That was it—the proof that the land had been uplifted. The great mountains had been built in small steps, as Lyell had written, and now Charles was an eyewitness to the process.

On a slope in the Andes, he found even more stunning testimony in a grove of fossilized trees at 7,000 feet. How could trees be sitting this high up, embedded in sandstone? Charles deciphered the geological explanation for this astonishing sight:

> I saw the spot where a cluster of fine trees had once waved their branches on the shores of the Atlantic, when that ocean (now driven back 700 miles) approached the base of the Andes . . . upright trees, had subsequently been let down to the depths of the ocean. There it was covered by sedimentary matter . . . but again the subterranean forces exerted their power, and I now beheld the bed of that sea forming a chain of mountains more than seven thousand feet in altitude and bearing trees that had once been buried in the seabed.

Land sinking, mountains rising—Charles now thought about everything in a dynamic geological perspective. On the island of San Lorenzo off the coast of Peru, he examined the shell beds that rose above the level of the sea. In a terrace, together with the shells, he was very curious to find cotton thread, plaited rush (braided seagrass), and the head of a stalk of Indian corn—signs of earlier human inhabitants. Charles deduced that the island had risen 85 feet since humans last lived there.

Geology dominated his thoughts. Off the coast of Peru, he began to think about the Pacific islands he was about to visit. One of the *Beagle*'s assignments was to take measurements around the picturesque coral islands and to see whether the rings of coral that encircled them sat on the rims of rising volcanic craters, as was then believed. Though he had not yet seen a coral island with his own eyes, Charles rethought the situation and came to the opposite conclusion. What if the mountains were actually sinking? Then the corals, which

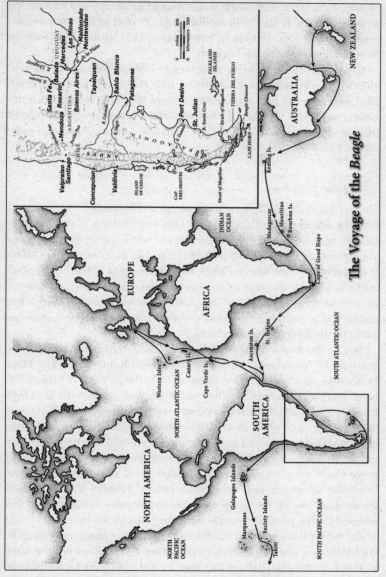

Figure 2.5 Map of the Voyage of the HMS *Beagle*, 1831–1836. Drawn by Leanne Olds.

required the light of shallow water, would grow upward around the sinking masses. If so, the beautiful rings of coral atolls were not sitting on crater rims but encircling sinking land forms. This was the first theory he could call his own.

Charles wrote to Henslow that he was looking forward to his next stop for two reasons: it would bring him that much nearer to England, and it would give him a chance to see an active volcano. But this time it would be the animals, not the landscape, that set his mind in motion. The *Beagle* sailed for the Galápagos Islands, some 600 miles off the coast (Figure 2.5).

Reptile paradise

He arrived in the Galápagos archipelago on September 15, 1835, well into the fourth year of the voyage. One might think that with these islands now inextricably linked with Darwin's name, they were the young naturalist's Eden. Far from it. In his diary of the first days there he wrote: "the stunted trees show little signs of life. — The black rocks heated by the rays of the vertical sun like a stove, give to the air a close & sultry feeling. The plants also smell unpleasantly. The country was compared to what we might imagine the uncultivated parts of the infernal regions to be."

But he did find a bay swimming with fish, sharks, and turtles, and the islands' "paradise for the whole family of Reptiles . . . The black lava rocks on the beach are frequented by large (2–3 ft.) most disgusting, clumsy lizards . . . They assuredly well become the land they inhabit." On a stroll, Charles encountered "two very large Tortoises (circumference of shell about 7 ft.). One was eating a Cactus & then quietly walked away . . . They were so heavy I could scarcely lift them off the ground.- Surrounded by the black lava, the leafless shrubs & large Cacti, they appeared most old-fashioned antediluvian animals; or rather inhabitants of some other planet" (Figure 2.6). He found great numbers of tortoises near the freshwater springs and was amused by the lines of animals marching to and fro.

On James Island, Charles collected all of the animals and plants he could. He was curious to decipher whether the plants were the same as those on the South American continent or peculiar to the islands. He also paid attention to the birds. The species of mockingbird on James Island looked different from those on two other islands. Mov-

Figure 2.6 A Galápagos tortoise. From *A Naturalist's Voyage Around the World: The Voyage of the H.M.S. Beagle,* by Charles Darwin (D. Appleton and Co., New York, 1890).

ing from island to island, the primary challenge was collection; identification would come later.

Charles did resolve the mystery of the marine iguanas and what they ate. A previous captain had concluded that they went out to sea to fish. But when Charles opened the stomachs of several animals, he found them packed with the seaweed that grew in thin layers on the submerged rock. Though the iguanas were hideous to his eye, Charles did admire their great swimming ability and diving endurance, noting that he believed these habits were unique among all lizards and strikingly different from those of the island's land iguanas.

After five weeks of hiking across sand that reached 137 degrees F, Charles and the *Beagle* sailed west.

Centers of creation, the mystery of mysteries

Charles, for once, enjoyed the long sail through the tropical seas to Tahiti. Midshipman King later recalled the pleasure Charles took in "pointing out to me as a youngster the delights of the tropical nights, with their balmy breezes eddying out of the sails above us, and the sea lighted up by the passage of the ship through the never-ending streams of phosphorescent animalculae."

The *Beagle* then went on to New Zealand, Australia, and the Cocos Islands. There Charles saw his first coral atolls, with their reefs encircling gorgeous blue lagoons. Wading among the corals, he immersed himself in the wonders of the reef and confirmed his suspicions about how such beautiful structures were built.

The *Beagle* then sailed on toward Africa, reaching it at the end of May 1836. At the Cape of Good Hope, Charles went ashore with the captain to call on the great astronomer Sir John Herschel, whose book Charles had read while studying at Cambridge. Herschel was keenly interested in geology and was also a close follower of and correspondent with Lyell. But he thought Lyell had missed the mark with his second volume of the *Principles of Geology.*

Lyell's new volume, a copy of which Charles received while at sea, focused on questions surrounding the appearance of species. The notion that species could change or "transmutate" had been floating around for decades in France and Britain but had not attracted general support. Part of the resistance was due to the lack of evidence, but a large part was that it conflicted with the creation of living things by a Creator, a view held by most of the establishment, including Lyell as well as Darwin's teachers. Although very knowledgeable about fossils, Lyell rejected evolution as an explanation for the appearance and disappearance of species. Like the other geologists of his day, Lyell adhered to the view of species as unchanging— with each kind specially created. He explained the succession of species in the fossil record as a succession of creations with species created "in succession at such times and in such places as to enable them to multiply and endure for an appointed period, and occupy an appointed place on the globe."

Herschel thought otherwise. If landscapes evolved, as Lyell had amply demonstrated, why not their inhabitants? Herschel saw a connection to "the mystery of mysteries"—the origin of new species. Whether he fully disclosed his thoughts to Charles is not clear. But it is clear that on the journey home, and thereafter, the mystery gripped Charles.

He had much to look forward to. Henslow had collected and published ten of his letters in a pamphlet, and Charles' sister wrote that his name was gaining much attention in England. Charles began to plot his return and to prioritize his work. Surrounded by reams of geological, zoological, and botanical notes and specimens, he began organizing them for publication. The final leg of the voyage, thanks to FitzRoy's obsessive chartmaking, was going to be longer than expected. Instead of moving up the west coast of Africa to Europe, the ship headed back to Brazil for one last check of some measurements. Charles, who never conquered his seasickness, wrote home: "I loathe, I abhor the sea."

But he was able to make good use of the extra time. He began to gather and to flesh out his ornithological notes and returned to the puzzle of the Galápagos birds. He concluded that the mockingbirds were closely allied in appearance to those of Chile, but he believed there was more to the story:

> I have specimens from four of the longer islands; the specimens from Chatham & Albemarle Isd. appear to be the same, but the other two different. *In each Isd. each kind is exclusively found* [emph. added]; habits of all are indistinguishable. When I recollect, the fact that from the form of the body, shape of scales & general size, the Spaniards can at once pronounce from which Isd. any tortoise may have been brought: —when I see these Islands in sight of each other and possessed of but a scanty stock of animals, tenanted by these birds but slightly differing in structure filling the same place in Nature, I must suspect they are only varieties. The only fact of a similar kind of which I am aware is the constant asserted difference between the wolf-like fox of East & West Falkland Isds.—*If there is the slightest foundation for these remarks, the zoology of Archipelagoes—will be well worth examining; for such facts would undermine the stability of species* [emph. added].

By the end of the voyage, Charles was already pondering that mystery of mysteries in a fresh light.

The mariner returns

It was a joyous and triumphant homecoming.

For five years he had been away from his friends, family, and mentors. His sisters were so relieved to see him home safe. And the Doctor —well, the Doctor was very proud. Charles had left, a directionless bug-catcher; he returned to the toasts of the cream of British scientific circles. He was most anxious to see Henslow again and to get his advice on what to do with his specimens.

The great Lyell wanted to meet him, and Charles soon was invited to a dinner at the London home of his geological hero. Lyell was transfixed by the tale of the Chilean earthquake, and he introduced Charles to the people who could help with the scientific analysis of his collections. The fossils, the birds, the plants, and even the iguanas found eager takers.

Charles considered writing a book about his long voyage. He loaned his diaries to his Wedgwood cousins, and they were very encouraging. FitzRoy planned to produce a three-volume "narrative" of the *Beagle*'s voyages, written by himself, a former captain, and Charles.

As Charles set to work writing his account, the experts were poring over his collections. The ornithologist John Gould, an accomplished naturalist and illustrator, quickly perceived that Charles'

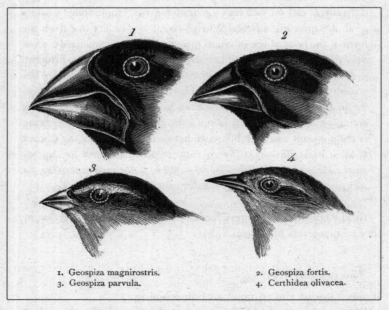

1. Geospiza magnirostris.
2. Geospiza fortis.
3. Geospiza parvula.
4. Certhidea olivacea.

Figure 2.7 Galápagos finches. From *A Naturalist's Voyage Around the World: The Voyage of the H.M.S. Beagle,* by Charles Darwin (D. Appleton and Co., New York, 1890).

Galápagos birds were closely related. What Charles thought were "gross-beaks" and "blackbirds" were actually finches. In just a few days' examination, Gould identified twelve (later revised to thirteen) species of ground finches—all entirely new species (Figure 2.7). And the "varieties" of mockingbirds included three distinct species. They were related to those in Chile, as Charles had surmised, but they were not identical to them.

Here was the crucial puzzle. How could Charles explain all of these new species, each specific to an island? The conditions on each island were not significantly different, so if each bird had been created to suit each island, why were they different? It was inescapable. The original birds that migrated to the islands had changed somehow and produced new species.

Charles knew it was difficult to explain, even more difficult to persuade others to believe, and as it violated the doctrine of the immutability of species and challenged creation-based explanations, he stood on dangerous territory. He was torn. He was eager for recognition and to climb the ranks of the scientific elite, but he knew that the "transmutation" of species was taboo. Neither his new boosters nor his Cambridge mentors would stand for such heresy.

He worked furiously on his *Journal* and tried to finesse the issues raised by the Galápagos animals:

[I]t never occurred to me, that the productions of islands only a few miles apart, and placed under the same physical conditions, would be dissimilar. I therefore did not attempt to make a series of specimens from the separate islands. It is the fate of every voyager when he has just discovered what object in any place is more particularly worth of his attention, to be hurried from it . . . It is clear, that if several islands have their peculiar species of the same genera, when these are placed together, they will have a wide range of character. *But there is not space in this work, to enter on this curious subject* [emph. added].

And so began the dodging game he would play for the next twenty years. When he wrote these lines, he was already convinced that species change, but he did not tip his hand. Indeed, when a young Alfred Russel Wallace read this passage, he saw the "mystery of mysteries" as an open question that the great Darwin had overlooked, and it spurred him to make his own voyages (Chapter 3).

Charles finished writing his *Journal of Researches into the Geology and Natural History of the Various Countries Visited by the H.M.S. Beagle Under the Command of Captain FitzRoy, R.N. from 1832 to 1836* (what we know as *The Voyage of the Beagle*) in seven months. (It did not appear for two years due to FitzRoy's delays in finishing his part.) Publicly, he would only go so far. Privately, he threw himself into the study of the "transmutation" of species.

Secret notebooks and a species theory

Charles was soon accepted into the elite scientific societies and hailed for his collections and his geological work. After the completion of his *Journal*, he began making notes on the transmutation of species.

He recalled the "ostriches" of South America. Early in the voyage, he had heard of a second, smaller form that lived in southern parts of Patagonia, beyond the Rio Negro. This Petise form was rare and Charles wanted one very badly, but they were difficult to spot and very wary, and he had no luck. While dining one night on what he casually thought was a juvenile ostrich, it dawned on him that he was actually consuming the elusive species. Panicked, he rescued some parts that had not yet been cooked or eaten. Years later, back in England, John Gould named the reassembled bird *Rhea darwinii*.

Now Charles was puzzled that the large and small rheas overlapped in their territories near the Rio Negro. Unlike the Galápagos birds, they had no boundary to separate them. The two species made him note: "one is urged to look to common parent?"

Charles opened up a new notebook ("B") and wrote on the title page in bold letters: "Zoonomia," picking up where his grandfather had left off nearly forty years earlier and a theme he first read as a struggling medical student. He jotted his thoughts down as they streamed out.

On page 15, he recalled the animals of Australia and scrawled:

Countries longest separated greatest differences—if separated from immens ages possibly two distinct type [sic], but each having its representatives—as in Australia. This presupposes time when no Mammalia existed; Australia Mamm. were produced from propagation from different set, as the rest of the world.

On page 20:

We may look at Megatherium, armadillos, and sloths as all offsprings of some still older type some of the branches dying out.

On page 21:

Organized beings represent a tree *irregularly branched* some branches far more branched—Hence Genera.—As many terminal buds dying as new ones generated.

On page 35:

If we grant similarity of animals in one country owing to springing from one branch. . .

And then, on page 36, after the declaration "I think," he drew a little diagram that represented a new system of natural history, a tree of life with ancestors at the bottom and their descendants at the top (Figure 2.8).

His jottings raced from topic to topic in zoology, geology, and anthropology. Each entry was a fragment of a much larger picture that was slowly taking form.

Life was a tree, with the branches and twigs connecting species, like relatives in a family pedigree. But what made the tree branch? Why were new forms arising and others dying out?

Through the next year, he read all sorts of books to try to answer the questions burning in his mind. On September 28, 1838, he opened Thomas Malthus' *Essay on the Principle of Populations*. Malthus proposed that there were checks on the growth of populations—disease, famine, and death—that prevented them from increasing at a geometric rate. He explained that there was a great overproduction of offspring in nature because of these checks. And what would sort out the survivors from the others? It was clear to Charles: the stronger, better-adapted ones. In his notebook he wrote: "One may say there is a force like a hundred thousand wedges trying [to] force every kind of adapted structure into the gaps in the economy of nature, or rather forming gaps by thrusting out weaker ones. The final cause of all this wedging, must be to sort out proper structure, and adapt it to change."

The result, Charles realized, would be the formation of new species.

His "species theory" was thus born, and it would grow and develop over the next few years. He quickly connected the analogy between the role of nature in shaping her species and the role of humans in shaping breeds. "It is a beautiful part of my theory that domesticated races of organisms are made by precisely the same means as species —but latter far more perfectly & infinitely slower." That natural process was to be called "natural selection."

Charles was also recalibrating the clock of life. Influenced by Herschel, the astronomer, who suggested that "the days of Creation" may correspond to "many thousand millions of years," Charles felt confident that the earth and life were much older than geologists had grasped.

But even as his certainty increased, all of these ideas remained private. Astronomy had broken down long-held prejudices and geology was beginning to make similar progress, but he knew that the origins of living things were seen by most as something altogether different and sacred. Charles no longer held that view; he did not think it made any sense. Within weeks of conceiving his species theory, he jotted in his "N" notebook: "we can allow satellites, planets, suns, universes, nay whole systems of universes to be governed by laws, but the smallest insect, we wish to be created at once by special act." His species theory was heresy to the dogma of special creation, and he knew that heretics were not treated well.

In 1839, his *Journal* appeared to great acclaim. Charles was becoming famous. And one day, a letter arrived from Potsdam. It was from the great Humboldt himself. Gushing with praise, Humboldt declared that his influence on Charles was "the greatest success that my humble work could bring."

Charles was thrilled and moved. He thanked his hero "for the great pleasure you have given me by your letter. That the author of those passages in the Personal Narrative, which I have read over and over again, & have copied out, that they might ever be present in my mind, should have so honoured me, is a gratification of a kind, which can but seldom happen to anyone."

It was too much of a risk to gamble his soaring reputation on his species theory.

The dear old philosopher

Charles had no time, nor the inclination, to return to his divinity studies. He was fully consumed by his science and felt that if he did not work very hard on the fruits of his voyage in the first few years home, it would overwhelm him. While his heart no longer belonged to his parsonage, he was eager to settle down and start a family. Just before he turned thirty, he married his first cousin Emma Wedgwood.

They had known each other all their lives. Charles confided to Emma where his thoughts were leading. Emma, a devout Christian, worried that Charles' heresies might preclude their eternal life together. It was a delicate balance. Emma knew that Charles was working on great ideas, but Charles was very mindful of Emma's concerns. He had yet another reason to keep his theory private.

In 1842, Charles distilled his notes and several years' thinking into a thirty-five-page sketch of his species theory. He drew on all he had learned during his voyage and since about the geographic distribution of animals, their variation, and the antiquity of fossils to put forth a new view of the origin of species, one that was completely and explicitly opposed to special creation. Charles explained his many reasons for concluding that species were not immutable. He summarized the evidence for his theory that species were descended and modified from earlier species, and he explained how that could occur through natural selection. He described a new view of nature: a war that involved incalculable waste, famine, death, and change.

Charles' criticism of creationism was blunt. Describing the very slight differences among three rhinoceros species from Java, Sumatra, and India, Charles found the idea that a Creator would make such similar but slightly different forms less than implausible; "Now the Creationist believes these three Rhinoceroses were created . . . as well as I can believe the planets revolve in their present course not from one law of gravity but from distinct volition of [the] Creator."

Two years later, he expanded this into an essay of 230 pages. Its Table of Contents is strikingly similar to that of *The Origin of Species*, which would not appear for another fifteen years, in 1859. Many of the well-known arguments and prose from his great book appear in these pages, including a form of the closing passage, which exalts the grandeur in this new, Darwinian view of life.

But Charles thought it was unwise, even reckless, to publish at the

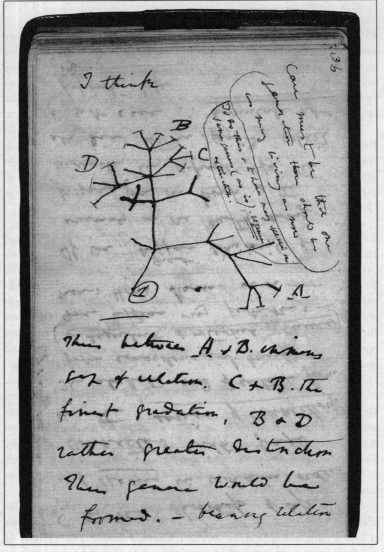

Figure 2.8 The tree of life. Page from notebook "B," where Darwin records his idea of life connected as in a tree, with ancestors at the bottom. Reproduced by kind permission of the Syndics of Cambridge University Library.

time. It would have meant breaking ranks with his teachers and supporters—Lyell, Henslow, and Sedgwick—as well as the rest of the scientific establishment and would be professional suicide. He also would certainly offend Captain FitzRoy, an ardent creationist, who welcomed him on his ship and looked after him for five years. He would, in time, share the essays with only a few trusted intimates—Lyell, the botanist Joseph Hooker, Thomas Huxley, and Emma. On July 5, 1844, he wrote his wife a note:

> . . . I have just finished my sketch of my species theory. If, as I believe, my theory in time be accepted even by one competent judge, it will be a considerable step for science.
>
> I therefore write this in case of my sudden death, as my most solemn and last request . . . that you will devote £400 to its publication, and further will yourself, or through Hensleigh [Emma's brother], take trouble in promoting it.

He then proceeded to work on barnacles and all sorts of topics in botany, zoology, and geology. Independently wealthy, thanks to his father and Emma's father, and ensconced at his manor in the village of Down, Charles' life revolved around his work, Emma, and their ten children (seven of whom lived to adulthood). A doting father, he often regaled his children with tales of his adventures on the *Beagle* and stories about his shipmates.

Charles' demeanor as a father and a husband was the same as that recalled by his shipmates, who never saw him out of temper or heard him say an unkind word about or to anyone in the course of their long journey. It was in admiration of these qualities, and his ability, that they had given him perhaps his most apt nickname: "the dear old Philosopher."

The Voyage of the Beagle would inspire another wave of naturalists, for whom Charles would play the same role that Humboldt had for him. And a letter from one of this new group of voyagers would finally, after twenty years, make him break his silence about the origin of species and publish his theory and greatest work.

Figure 3.1 Sketch salvaged from fire and shipwreck of the *Helen*. This drawing of an Amazonian angelfish was one of the few sketches Wallace managed to save out of all of his notes and specimens on his doomed voyage home. It displays one of the important talents for naturalists before the age of photography —that of being a good artist. Drawing from *My Life*, by Alfred Russel Wallace (New York: Dodd, Mead, and Co.).

∽ 3 ∽
Drawing a Line between Monkeys and Kangaroos

All truths are easy to understand once they
are discovered; the point is to discover them.
— *Galileo Galilei*

IT WAS TIME TO PACK UP and go home.

Alfred Wallace was 2,000 miles upriver from the Atlantic Ocean, on the Rio dos Uaupés tributary of the Amazon—farther than any European had ever gone. Since arriving with Henry Walter Bates in May 1848, he had spent nearly four years exploring and collecting, the last two and a half years on his own. But he had been laid up for the last three months with yellow fever and was too exhausted to continue. His younger brother Herbert, who followed him to Brazil and accompanied him up to the Rio Negro, had long before turned back. Unbeknownst to Wallace, Herbert had been stricken with yellow fever and died before he could board a boat to England.

Wallace had accumulated a large menagerie of animals—monkeys, macaws, parrots, and a toucan—that he hoped to take all the way to the London Zoo, and their upkeep was literally killing him. He also had a couple of years' worth of specimens, both with him and stored downriver, that he had not yet been able to ship to England for sale.

Wallace began to dream of green fields, neat gardens, bread and butter, and other comforts of home. He boarded the brig *Helen* with thirty-four live animals and many boxes of specimens and notes and set sail for England.

"I'm afraid the ship's on fire; come and see what you think of it." Just after breakfast, three weeks out of port and somewhere east of Bermuda, the captain was concerned enough to visit Wallace in his cabin. And rightfully so. Smoke was pouring out of the hold.

The crew tried but could not douse the smoldering blaze. The captain ordered down the lifeboats. Wallace, still weak, looked on as if it were a feverish dream. It wasn't. He reentered his hot, smoky cabin and salvaged a small tin box and threw in some drawings (Figure 3.1), some notes, and a diary. He grabbed a line to lower himself into a lifeboat, slipped, and seared his hands on the rope. His pain was compounded when his injured hands hit the salt water. Once in the lifeboat, he discovered it was leaking.

Wallace watched his animals perish, then the *Helen*, along with all of his specimens.

There he was, lying on his back in a leaky lifeboat in the middle of the Atlantic. His loss had not yet fully dawned upon him. He was optimistic about being rescued and enjoyed the dolphins frolicking around him. But the wind changed, and day after day passed in the open boats. Wallace was blistered with sunburn, parched with thirst, soaked by sea spray, exhausted by constantly bailing water, and near starvation. At last, on the tenth day, they were picked up.

That first night aboard his rescue ship, the *Jordeson*, Wallace could not sleep. He wrote: "It was now, when the danger appeared past, that I began to feel fully the greatness of my loss . . . How many weary days and weeks had I passed, upheld by the fond hope of bringing home many new and beautiful forms from those wild regions; every one of which would be endeared to me by the recollections they would call up, which should prove that I had not wasted the advantages I had enjoyed and would give me occupation and amusement for many years to come! And now everything was gone, and I had not one specimen to illustrate the unknown lands I had trod."

Actually, the danger was not yet past. The *Jordeson* was twice hit by heavy storms. One tore the sails and sent a wave crashing through Wallace's skylight, soaking him while he slept. The second hit in the English Channel, when they were almost home. It sank many ships and put four feet of water in the *Jordeson's* hold.

Wallace swore to himself "fifty times" on the voyage home that "if

I once reached England, never to trust myself more on the ocean." Had he kept that promise, his story would end here, and few would have ever heard of Alfred Wallace again.

He broke his vow within days.

Where to next?

Thank goodness he chose Samuel Stevens, the agent who sold what Wallace had managed to ship to England before his calamity. Stevens met Wallace in London and bought him a new suit, and Stevens' mother fed him until his strength returned. Moreover, Stevens had had the foresight to insure Wallace's collections for 200 pounds. It was not as much money as Wallace had hoped to gain by selling his Amazonian treasure, but it was enough to keep him from begging.

The loss of his specimens, rather than deterring Wallace from further adventures, only stoked his determination. His voyage was not finished. His lust for exploration and collecting was not satisfied, nor was his interest in the origin of species. That mystery was still unsolved, as far as the scientific world knew in 1852. Though Darwin had written his essays a decade earlier, their contents were known only to a few, and Wallace was not one of them. Now thirty years old, he was, unlike Darwin, not yet ready to settle down.

He began to ponder his next journey. The big question was "where to go?" He had both practical and scientific matters to consider. He had heard the fiery young zoologist Thomas Huxley say, "Science in England does everything—except pay." For working-class, self-made men like Huxley and Wallace, it was a painful truth. Wallace had to collect quarry that would fetch good prices, so he ruled out a return to the Amazon. His former companion, Henry Walter Bates, was still there and had that territory covered. No, it had to be someplace new.

He kept thinking about the Malay Archipelago, the vast group of islands between Southeast Asia and Australia. Other than those on the island of Java, the animals and plants of the region had not been explored. Enough fragments of natural history were emerging from the Dutch settlements there to convince Wallace that it offered both rich pickings and good facilities for a traveler. The islands span more than 4,000 miles from east to west and 1,300 miles from north to south, an area almost the size of the entire continent of South America (Figure 3.2). Many of the islands are volcanic (one, Krakatoa, would nearly vaporize in 1883 in an enormous eruption that altered the planet's climate). Covered in tropical forest, the islands appeared similar, but some held different treasures, and discovering and explaining the differences would put Wallace, literally, on the map.

On the hunt again

It was a much longer journey to the Far East than to Brazil. Arriving in April 1854, Wallace set out to explore the country. He encountered altogether different treasures, and dangers, than those on the Amazon. For example, the island of Singapore was an excellent collecting ground for insects. There were, however, some drawbacks: "Here and there, too, were tiger pits, carefully covered over with sticks

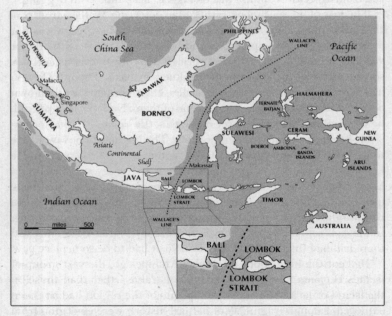

Figure 3.2 The Malay Archipelago and the Wallace Line. Wallace spent eight years traveling among these islands. He discovered that the narrow strait between Bali and Lombok marked a boundary between Asiatic and Australia-type fauna. Bali was once connected to the Asiatic continental shelf, but not to Lombok. The boundary, dubbed the "Wallace Line," extends throughout the archipelago as shown. Map by Leanne Olds.

and leaves, and so well concealed, that in several cases I had a narrow escape from falling into them . . . fifteen or twenty feet deep, so it would be almost impossible for a person unassisted to get out of one."

There were tigers roaming about Singapore: they took on average one resident a day. Wallace occasionally heard their roars and, in typical British understatement, he noted, "It was rather nervous work hunting for insects . . . when one of these savage animals might be lurking close by."

The natives were rumored to be just as dangerous. A friend, seeing that Wallace had his mattress on the floor of his bamboo house, told him it was very dangerous "as there were many bad people about, who might come at night and push their spears up through me from below, so he kindly lent me a sofa to sleep on, which, however, I never used as it is too hot in this country."

Unflustered by such concerns, Wallace had a daily routine. Up at 5:30 A.M., he started with a cold bath and hot coffee. He sorted out the previous day's collection, then set out again into the forest with his gear. He carried a net, a large collecting box hung on a strap over his shoulder, pliers for handling bees and wasps, and two sizes of specimen bottles for large and small insects, attached by strings around his neck and plugged with corks. And on some days he carried a rifle.

To preserve his specimens and skins, he relied on the locally brewed arrack. About 70 percent alcohol by volume, arrack was distilled from various fermented fruits, grains, sugarcane, or coconut sap. There was a flourishing trade in the beverage, so Wallace was usually well stocked. However, the natives' taste for the drink was very keen, so casks often went missing from his house or field camps. Wallace then developed the defensive strategy of putting dead snakes and lizards into his casks, but even this maneuver did not deter many from drinking his supplies.

The natives could not understand why all the animals, birds, insects, and plants were preserved so carefully (using up good arrack!). Wallace told them that people in his country would go to look at them, but this made no sense. Surely there must be better things to see in "Ung-lung" (as one Wanumbai tribesman pronounced "England")? For their part, some of the tribes Wallace met seemed to enjoy their own hobby of collecting: The headhunting Dyaks kept bundles of their enemies' heads suspended from the ceilings of their longhouses.

Despite some reputation for ferocity, the native tribesmen shared their knowledge of the forest with Wallace and helped him find what he was after. He stalked its most beautiful and prized treasures—orangutans, monkeys, spectacular birds of paradise, and enormous brilliant butterflies. He mused:

> Nature seems to have taken every precaution that these, her choicest treasures, may not lose value by being too easily obtained. First, we find an open harbourless, inhospitable coast, exposed to the full swell of the Pacific Ocean; next, a rugged and mountainous country, covered with dense forests, offering in its swamps and precipices and serrated ridges an almost impossible barrier to the central regions; and lastly, a race of the most savage and ruthless character.

No matter how many years he spent in the forest, the thrill of capturing something new never diminished:

> I had seen sitting on a leaf out of reach, an immense butterfly of a dark colour marked with white and yellow spots . . . I at once saw that it was a female of a new species of Ornithoptera or "bird-winged butterfly," the pride of the Eastern tropics. I was very anxious to get it and to find the male, which in this genus is always of extreme beauty. During the two succeeding months I only saw it once again . . . I had begun to despair of ever getting a specimen . . . till one day . . . I found a beautiful shrub . . . and saw one of these noble insects hovering over it, but it was too quick for me, and flew away. The next day I went again to the same shrub and succeeded in catching a female, and the day after a fine male . . . more than seven inches across the wings, which are velvety black and fiery orange, the latter color replacing the green of the allied species. The beauty and brilliancy of this insect are indescribable . . . on taking it out of my net and opening the glorious wings, my heart began to beat violently, the blood rushed to my head, and I felt much more like fainting than I have done when in apprehension of immediate death. I had a headache the rest of the day. (Figure 3.3).

Thinking out loud

That pounding head was doing more than fawning over butterflies. Wallace was paying close attention to the diversity of species he found,

Figure 3.3 The Golden Birdwing butterfly. The large bird-wing butterflies of Indonesia were highly sought by collectors. Their varieties
from island to island prompted Wallace to consider the nature of species,
and to doubt that special creation would explain such variety. Wallace
discovered this form (*Ornithoptera croesus lydius*) on the island of Batjan.
Photo by Barbara Strnadova.

the variety among the individuals of each species, and *where he found
them*. These were not only the practical concerns of a paid collector
but also the catalysts of his transformation into a scientist.

While Darwin was keeping his silence about evolution, Wallace
was thinking out loud, putting his thoughts on paper and dashing
them off to magazines and journals in England. Some of them were
short field notes, but others revealed bigger ideas. Wallace was puz-
zling over some of the same facts and observations as Darwin had
and reaching some remarkably similar conclusions. But Wallace had
none of the concerns that had restrained Darwin. He had a reputation to make and nothing to lose.

In 1855, while waiting out the wet season in Sarawak, on Borneo,
Wallace wove together threads of geology and natural history to propose a new law: *"Every species has come into existence coincident both in
space and time with a pre-existing closely allied species."*

Wallace thought that species were connected like "a branching
tree." He was proposing that new species come from old species the
way new twigs grow from older branches. The idea may not sound

too dangerous, but it was very bold. It targeted the well-accepted doctrine of special creation—that each species was specially created, in one moment, to fit the land it inhabited. Moreover, he was using some of the very arguments that Darwin had agonized over for almost two decades but had not yet published.

Wallace embraced the growing picture from geology of a changing earth and from the fossil record of the obvious changes in life. He simply extrapolated that what was true of the past must be true of the present—"that the present geographical distribution of life upon the earth must be the result of all the previous changes, both of the surface of the earth itself and of its inhabitants." In short, the earth and life evolve together. Folks were starting to get used to the idea of the earth changing; they didn't at all like the idea of life evolving.

Wallace supported his "Sarawak Law" with all sorts of observations on the distribution of species, especially those on islands. Take the Galápagos, for example: "which contain little groups of plants and animals peculiar to themselves, but most nearly allied to those of South America, have not hitherto received any, even a conjectural explanation." He was taking a swipe at Darwin, who had hitherto dodged the subject. Wallace continued: "They must have been first peopled, like other newly formed islands, by the action of winds and currents, and at a period sufficiently remote to have had the original species die out, and the modified prototypes only remain." Translation: There is no finch species in South America identical to those on the Galápagos, but there are such close allies that South American finches must have colonized the islands.

Wallace pointed out that families of birds, butterflies, and various plants are confined to certain regions. He had noticed in the Amazon that some species of monkeys were confined to one side of the river. "They could not be as they are, had no law regulated their creation and dispersion." By "dispersion," he meant that the extent to which a species could spread out over the land was constrained by the features of the land—rivers, mountain ranges, and so forth.

Almost no one read or noticed his paper when it first appeared. Wallace heard nothing from England except for some grumblings that he should focus on collecting, not theorizing.

He did hear from his old friend Bates, who, although camped in the Upper Amazon, managed to get a copy of the journal. Bates heartily congratulated Wallace on his idea, which he thought was "like truth itself, so simple and obvious that those who read and understand it will be struck by its simplicity."

Drawing a line

Wallace hopped from one island to another quite often. He made ninety-six journeys, covering about 14,000 miles, and visited some of the same islands several times over the span of eight years. He had to be flexible. Often the availability of a boat determined his path. He tried several times to get from Singapore to Makassar, on the island of Sulawesi, without any luck. But one day in May 1856 he did find a Chinese schooner headed to Bali; he had no intention of going there, but he figured he could find a way from there to Lombok and then on to Makassar. This accidental detour would lead Wallace to the most important discovery of his expedition.

On Bali, Wallace found the same kinds of birds as on the other islands he had visited—a weaver, a woodpecker, a thrush, a starling—nothing too exciting. Then, "crossing over to Lombok, separated from Bali by a strait less than twenty miles wide, I naturally expected to meet with some of these birds again; but during a stay there of three months I never saw one of them." Instead, he found a completely different assortment—white cockatoos, three species of honey-suckers, a loud bird the locals called a Quaich-Quaich, and a really strange bird called a megapode ("big foot"), which used its big feet to make very large mounds for its eggs. None of these groups was known on the western islands of Java, Sumatra, Malaysia, or Borneo.

Here was a puzzle. What prevented the spread of these species from island to island? Surely, birds could cover a twenty-mile strait with little trouble.

Wallace described the mystery in a letter to Bates. He theorized that there was some kind of invisible "boundary line" between Bali and Lombok (see map, Figure 3.2). Traveling farther east to Flores, Timor, the Aru Islands, and New Guinea, the change in bird life was very clear. All of the families of birds that were common on Sumatra, Java, and Borneo were not found on Aru, New Guinea, and Australia, and vice versa.

The differences in mammals among the western and eastern islands were just as striking. On the large western islands there were monkeys, tigers, and rhinoceroses. But on Aru there were no primates or carnivores; all the native mammals were *marsupials*—kangaroos and cuscus.

That line between Bali and Lombok was real, and it signified something very profound to Wallace. He put his thoughts on paper again:

Let us now examine if the theories of modern naturalists will explain the phenomena of the Aru and New Guinea fauna . . . How do we account for the places where they came into existence? Why are not the same species found in the same climates all over the world? The general explanation given is, that as the ancient species became extinct, new ones were created in each country or district, adapted to the physical conditions of that district.

By "created," Wallace meant specially created by a Creator. But, Wallace pointed out, this "theory" would make us expect to find similar animals in countries with similar climates and dissimilar animals in countries with dissimilar climates—which is not at all what he saw.

Comparing Borneo (in the west) and New Guinea (in the east), he wrote: "It would be difficult to point out two countries more exactly resembling each other in climate and physical features." But their birds and mammals were entirely different.

Now compare New Guinea and Australia: "We can scarcely find a stronger contrast than in their physical conditions . . . one enjoying perpetual moisture, the other with alternatives of drought." Wallace reasoned, "If kangaroos are especially adapted to the dry plains and open woods of Australia, there must be some other reason for their introduction into the dense damp forests of New Guinea, and we can hardly imagine that the great variety of monkeys, squirrels, of Insectivores, and Felidae [cats], were created in Borneo because the country was adapted to them, and not one single species given to another country exactly similar and at no great distance." In the tropical forests of the eastern islands, tree kangaroos occupied the habitat occupied by monkeys in the west.

The reason must be that "some other law has regulated the distribution of existing species." That law, Wallace suggested, was the "Sarawak Law" he had proposed two years earlier. Again he relied on geology to make his case. He surmised that New Guinea, Australia, and Aru must have been connected at some time in the past and so shared similar sets of birds and mammals. And the western islands?

Wallace deduced they had once been part of Asia and so shared the fauna tropical of Asia—monkeys, tigers, and the like.

Wallace was right. The distance between Bali and Lombok is short, but the ocean separating them was later discovered to be very deep. Bali lies just on the edge of the continental shelf, while Lombok lies just off it (Figure 3.2). Bali was once connected to the other western islands but never to Lombok. It wasn't simply a matter of flying the twenty miles to the next island. For millions of years the separation was much greater, and the animal life adapted to the conditions peculiar to each island. Today, the islands are close together, but they are, geologically speaking, "new neighbors."

Wallace had linked the question of the origin of species to how species were distributed and had defined a dividing line between the fauna of Asia and Australia. His discovery would forever be known as the "Wallace Line" (Figure 3.2), and Wallace himself as the founder of biogeography.

And, finally, Wallace was getting some attention from England. He had struck up a correspondence with Darwin in which he lamented that his Sarawak Law had not received any attention, or even opposition. In May 1857, Darwin replied, "I agree to the truth of almost every word of your paper; & I daresay that you will agree with me that it is very rare to find oneself agreeing pretty closely with any theoretical paper." He went on to explain that he was now in his twentieth year of examining the question of how species differ and well into the writing of a large book, which he did not expect to finish for two years. This was a bit of his marking his territory, some notice from the senior naturalist that he had been thinking about these matters for a long time and would, in due course, disclose his full thoughts. But it was perhaps Darwin who should have been put on notice, for Wallace was getting closer.

Survival of the fittest

For Wallace the question was then, not if species evolved, but how? Baking in a malarial fever on the volcanic island of Ternate in early 1858, the answers came to him.

Alternating between hot and cold fits, Wallace had nothing to do but "to think over subjects then particularly interesting to me." Wrapped in a blanket on an 88-degree day, he thought of Malthus' essay on population, which he had read some years earlier. It occurred to him that the diseases, accidents, and famine that check the growth of human populations act on animals, too. He thought about breeding, how animals bred much more rapidly than humans and, if left unchecked, would overcrowd the world very quickly. But all of his experience revealed that animal populations were limited. "The life of wild animals," Wallace concluded, "*is a struggle for existence.* The full exertion of all their faculties and all their energies is required to preserve their own existence and provide for that of their infant offspring." Finding food and escaping danger ruled animals' lives—and the weakest would be weeded out.

Wallace the great collector was intimate with the variety of individuals of a species; he wrote: "Perhaps all the variations . . . must have some definite effect, however slight, in the habits of or capacities of the individuals . . . a variety having slightly increased powers . . . most inevitably acquire a superiority in numbers."

Bingo. He had figured it out—or else he was out of his mind. Wallace had to wait for his fever to abate before he could make any notes. Then he wrote the paper out in full in just a few nights.

He called it "On the Tendency of Varieties to Depart Indefinitely From the Original Type." Later he would refer to his idea as "survival of the fittest," a phrase borrowed from the social scientist Herbert Spencer. Wallace's paper was just a sketch, conceived in a dilapidated house on an earthquake-ravaged island during bouts of fever, 10,000 miles from the center of science in England. Wallace did not send it directly to a journal; he wanted others to look at it first.

He sent it to—whom else?—Darwin.

This time, he would not go unnoticed.

Priority and posterity

When Darwin received Wallace's paper sometime in June 1858, he was shocked. He should not have been, had he been paying closer attention to all of Wallace's previous dispatches. Nevertheless, Darwin, now sixteen years past the first version of his "Essay" on species formation, "feared that all of my originality, whatever it may amount to, will be smashed."

What happened thereafter is still a subject of debate among scholars. The facts are that Wallace had asked Darwin to forward the manuscript to the geologist Sir Charles Lyell, which he did. Lyell and J. D. Hooker, the eminent botanist, were intimates of Darwin to whom he had divulged his theory of natural selection and much of the supporting argument. Lyell and Hooker took the initiative to arrange for Wallace's paper, and a brief sketch from Darwin on his theory, to be read together at an upcoming meeting of the Linnean Society and to be published together.

Was Wallace robbed of his individual right to glory? Was the joint publication fair? (Wallace was not informed of it until after the fact.) On the other hand, it was Darwin who coined the term "natural selection," and he had shared his 1842 sketch, at least privately, with other scientists.

It is true that today, Darwin's name and works are far better known than Wallace's. But consider Wallace's perspective on the matter. He always, for the rest of his life, deferred to Darwin. The year after *The Origin* appeared he wrote to Bates: "I know not how or to whom to express my admiration of Darwin's book . . . I do honestly believe that with however much patience I had worked up & experimented on the subject I could never have approached the completeness of the book, —its overwhelming argument, & its admirable tone & spirit . . . Mr. Darwin has created a new science & a new Philosophy, & I believe that never has such a complete illustration of a branch of human knowledge, been due to the labours and researches of a single man."

Wallace always referred to the "Darwinian theory," and he later dedicated his major book about his travels, *The Malay Archipelago* (1869), to "Charles Darwin, author of *The Origin of Species*, not only as a token of personal esteem and friendship but also to express my deep admiration for his genius and his works." In his autobiography, *My Life* (1905), he devoted a full chapter to his friendship with Darwin, with not a word of regret, envy, or resentment.

Perhaps for Wallace it was simply a matter of being accepted. He was, up until 1858, outside the circle of eminent scientists who led the new revolution in thought. When he heard that Lyell and Hooker had made complimentary remarks about his paper, he wrote to his oldest friend and school-fellow that "I am a *little* proud." Wallace did not need or seek to be the center of the circle: he just wanted to be let inside.

That, and more, he surely earned.

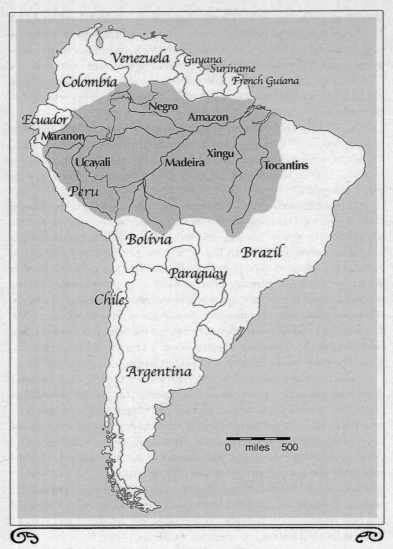

Figure 4.1 **The Massive Amazon River System.** The main river and its tributaries span more than 15,000 miles. Henry Walter Bates spent most of his eleven years in the Amazon on the main river, while Wallace ventured far up the Rio Negro. Map by Leanne Olds.

∾ 4 ∾
Life Imitates Life

> A river always leads to some inhabited place. If we do not meet
> with agreeable things, we shall at least meet with something new.
> — *Cacambo in* Candide, *Voltaire (1759)*

"IT WAS THE BEST of times, it was the worst of times . . ." So begins
Dickens' famous novel of 1859, *A Tale of Two Cities*. Returning from
the Amazon that same year, Henry Walter Bates could have begun
his tale with the same words.

The best of times was certainly daily life in a naturalist's paradise:

I rose generally with the sun, when the grassy streets were wet with
dew, and walked down to the river to bathe: five or six hours every
morning were spent in collecting in the forest . . . the hot hours of
the afternoon . . . and the rainy days were occupied in preparing and
ticketing the specimens, making notes, dissecting, and drawing. I frequently had short rambles by water in a small montaria [type of
canoe] . . . the neighborhood yielded me . . . an uninterrupted succession of new and different forms in the different classes of the animal kingdom, but especially insects.

The worst of times was perhaps the first year on his own, upriver,
after separating from Wallace in March 1850:

Twelve months elapsed without letters or remittances. Toward the
end of this time my clothes had worn to rags: I was barefoot, a great

inconvenience in tropical forests, notwithstanding statement to the contrary that have been published by travelers; my servant ran away, and I was robbed of nearly all my money.

Broke, lonely, unsure of his prospects, and pressed to return home to the family hosiery business, Bates traveled 1,400 miles downriver to the port town of Pará, intending to find a boat to take him home. Yellow fever had ravaged the place, however, and Bates too was stricken.

But he did not go home. He turned around and spent eight more years in the Amazon, eleven years in all. *Eleven years*. Why did he stay, and for so long? How did he bear it?

The answer to the first question is that Bates received a timely dose of new funds and a letter from his agent in London saying that his specimens were being very well received. One new butterfly species, *Callithea batesii*, had been named after him. He changed his mind and renewed his determination to go far upriver, as he had originally planned.

The length of his adventure was dictated by the massive scope of the Amazon basin. With more than a thousand tributaries draining about 2.7 million square miles, it is the largest of any river system (Figure 4.1). Bates roamed up and down about 2,000 miles of the main river in the course of these years (the ten largest rivers in the system span 15,000 miles). Travel was almost exclusively by water and was slow—very slow. Powered by only paddles or sails, pounded by storms and rain, and subject to changing winds (or no wind at all), Bates was often a passenger in some small trading vessel plying the river or in a canoe belonging to one of the many local tribes. Here's a scene from a typical crossing Bates made so that he could hunt for a monkey on the opposite bank:

We were about twenty persons in all, and the boat was an old rickety affair . . . In addition to the human freight we took three sheep with us . . . Ten Indian paddlers carried us quickly across . . . When about half-way, the sheep in moving about, kicked a hole in the bottom of the boat. The passengers took the matter very coolly, although the water spouted up alarmingly, and I thought we should inevitably be swamped. Captain Antonio took his socks off to stop the leak, inviting me . . . to do the same, whilst two Indians baled out the water . . . We thus managed to keep afloat.

By managing only a few miles or so at a stretch, the branches and twigs of the Amazon system seemed endless. Nevertheless, Bates wanted to see all of its treasures. That passion and determination, and the rewards of dashing into the forest at nearly every bend, seem to have offset the unrelenting heat, malaria, yellow fever, fire ants, biting flies, and his intense loneliness.

Those rewards were many—river dolphins, anteaters, frigate birds, anacondas, hummingbirds, bird-eating spiders, all sorts of monkeys, jaguars, caimans, blue hyacinthine macaws, parrots, eagles, five species of toucans, and butterflies—flocks of butterflies. Bates collected 14,712 animal species in all, of which more than 8,000 were new to science.

Eventually, the grueling work, bad and insufficient food, and overall deterioration of his health convinced Bates to return to England. The parting was bittersweet:

On the evening of the third of June [1859], I took a last view of the glorious forest for which I had so much love, and to explore which I had devoted so many years. The saddest hours I ever recollect to have spent were those of the succeeding night, when the mameluco pilot left us free of the shoals and out of sight of land . . . I felt that the last link which connected me with the land of so many pleasing recollections was broken . . .

Recollections of English climate, scenery, and modes of life came to me with a vividness I had never before experienced during the eleven years of my absence. Pictures of startling clearness rose up of the gloomy winters, the long grey twilights, murky atmosphere, elongated shadows, chilly springs, and sloppy summers . . . To live again amidst these dull scenes I was quitting a country of perpetual summer . . . It was natural to feel a little dismayed at the prospect of so great a change.

When Bates returned in the summer of 1859, his timing was quite fortunate. Within months, Darwin's *Origin of Species* appeared and gave Bates a concrete framework for thinking about all he had seen and collected.

The butterflies of Ega

No group of animals made a greater impact on Bates than the butterflies. They were, of course, highly prized in England for their beauty. Since Bates made his living by selling specimens, he paid careful attention to the varieties in each locale he visited.

The variety was overwhelming. In the vicinity of Ega alone, in the Upper Amazon where Bates spent more than four years, he found 550 distinct species of butterflies. This figure towers over the just 66 species in all of Britain and the 300 or so in all of Europe.

The butterflies at Ega and throughout the Amazon presented several puzzles to Bates' expert eye. For example, despite years of experience, he could never tell some species of *Leptalidae* from those of *Heliconidae* while they were in flight. Their wing markings were very similar, and they flew together in the same parts of the forest. Only on close inspection after capture could Bates identify which was which. The different varieties of just one species, *Leptalis theonoë*, resembled several different species of *Ithomia* butterflies.

Bates was very careful to determine where certain varieties were present or absent. He noticed that none of the particular *Leptalidae* variety that resembled a particular *Ithomia* species was found in any other district or country. The "counterfeits" were only passing themselves off where the real species existed in abundance. He called this phenomenon "mimetic analogy," or mimicry.

When Bates read *The Origin of Species*, he was one of the few immediate adherents. He, too, was an eyewitness to the war of nature and all of the strategies that animals used in battle. As he contemplated his butterflies, he realized that mimicry was evidence for the process of natural selection. He struck up a correspondence with Darwin in 1860, just as the firestorm of controversy was erupting over the great book. Bates wrote: "I think I have got a glimpse into the laboratory where Nature manufactures her new species."

Darwin was thrilled. Bates had no formal scientific position and, for the first three years after returning from the Amazon, he lived in Leicester with his family. He was feeling a bit discouraged at not being part of the scientific establishment.

Darwin rooted him on, urging him to present his work to the most important bodies, to publish in the most influential journals, and to write an account of his journey as a travel narrative, just as Darwin had done for his *Beagle* voyage. Bates gobbled up Darwin's advice. He often visited Darwin at his home in Downe, a privilege extended to very few. (Wallace was another, and all three explorers enjoyed a weekend together a bit after Wallace's return to England.) Darwin enjoyed Bates' company and admired his character. Theirs was a warm, symbiotic friendship.

One of the most remarkable and admirable papers

Bates set to work on both a formal scientific description of his collections and a book on his travels. Both were massive tasks. Bates later remarked that he would rather spend another eleven years in the jungle than go through the ordeal of writing another book.

But the same discipline that brought him success in the jungle brought him success as a scientist and author. His most important paper, with the misleadingly dull title "Contribution to an Insect Fauna of the Amazon Valley, Lepidoptera: Heliconidae," laid out the evidence and a mechanistic explanation for the phenomenon of mimicry.

Bates noted that several species of *Dioptis*, a genus of moths, also mimic species or local varieties of *Ithomia* butterflies. He explained that a series of mimetic relationships also occurred in the Old World between Asiatic and African *Danaidae* butterflies and species of other families of butterflies and moths. Most important, he underscored that no instance was known in these families of a tropical species of one hemisphere counterfeiting a form belonging to the other. In other words, these were not accidental resemblances of butterflies with different ranges; the mimicry occurred among species found in the same area (Figure 4.2).

Furthermore, Bates knew at first hand that mimicry occurs among other insects. Along the banks of the Amazon, he found parasitic bees and flies that mimic the forms of nest-building bees and live, "all expenses paid," in their nests. He found a cricket that was a good imitation of a Tiger Beetle and was always found on the trees it frequented. The most striking example of imitation was a very large caterpillar; when stretched in the foliage of a tree, it startled him by its resemblance to a small snake. The caterpillar had black spots on

Figure 4.2 Mimicry in butterflies. This is an original plate from Bates' 1862 paper reporting the discovery of mimicry. The butterfly at the center (5) is *Leptalis nehemia*, the typical butterfly of the family. The other Leptalis butterflies (1–8) deviate greatly from this pattern as they are mimics of other species. Each pair—3/3a, 4/4a, 6/6a, 7/7a, 8/8a— illustrates mimicry between Leptalis and species of other families.

segments of its head, which, when expanded, resembled the head of a pit viper (Figure 4.3). When Bates carried the specimen into the village, it frightened everyone who saw it.

Bates saw these phenomena in a Darwinian light. He proposed that the specific mimetic forms of insects were adaptations. He witnessed, in the Amazonian jungle, that every species maintained its existence by virtue of some traits that enabled it to withstand "the battle of life." He knew hundreds of examples of how animals conceal themselves from their enemies. Clearly, one species disguising itself as another was one of these strategies. In Bates' words, "the adaptive resemblance of an otherwise defenseless species to one whose flourishing race shows that it enjoys particular advantages."

The advantages of mimicking a poisonous snake are obvious, but what advantages did the *Heliconidae* butterflies possess that made them so abundant and the objects of imitation? It was not obvious what helped them escape the many insect eaters of the forest. Bates, though, had a good and ultimately correct idea. He knew too well that some butterflies secrete foul-smelling fluids and gases when handled. He noticed that when he laid such specimens out to dry, the various jungle vermin were less likely to carry them off. Bates also never saw the flocks of slow-flying *Heliconidae* pursued by birds or dragonflies, for which they would be easy prey, nor when resting were they attacked by lizards or predacious flies that pounded on other butterflies. Bates surmised that the *Heliconidae* must be unpalatable and that other species disguise themselves by mimicking their wing patterns; though palatable, they are protected from predators.

Bates saw the origin of mimicry, in general terms, as the same process that involves the origin of all species and adaptations. The case of *Leptalis theonoë* was most telling. The form of the species in each district depended on the form and colors of the *Ithomia* butterflies in each district, which varied from place to place. Bates asked, "How are local races formed out of the natural variations of a species?"

> The explanation of this seems to be quite clear on the theory of natural selection, as recently expounded by Mr. Darwin in the "Origin of Species" . . . If a mimetic species varies, some of its varieties must be more and some less faithful imitations of the object mimicked. According, therefore, to the closeness of its persecution by enemies, who seek the imitator, but avoid the imitated, will be its tendency to

become an exact counterfeit,—the less perfect degrees of resemblance being, generation after generation, eliminated, and only the others left to propagate their kind . . . To exist at all in a given locality, our *Leptalis theonoë* must wear a certain dress and those of its varieties that do not come up to the mark are rigidly sacrificed . . . I believe the case offers a most beautiful proof of the theory of natural selection.

So did Darwin. He called it "one of the most remarkable and admirable papers I ever read in my life," and he assured Bates "that it will have lasting value."

For Darwin and other proponents of natural selection, Bates' work was a powerful display of the engine of evolution. Darwin's evidence in *The Origin of Species* had relied heavily on the analogy of natural selection to the domestication of animals. Now, he had a rich, independent line of evidence from Nature.

Concerned that some might overlook Bates' work, Darwin wrote an account of it in the *Natural History Review*, pointing squarely at creationists and the doctrine of the immutability of species. He did not think that many naturalists could believe that all of the varieties Bates had discovered were specially created for each district—"turned out all ready made, almost as a manufacturer turns out toys according to the temporary demand of the market." No, Darwin said, through Bates' description "we feel to be as near witnesses, as we can ever hope to be, of the creation of a new species on this earth."

Argument and evidence

Mimicry became a focal point of the debate between proponents and opponents of the theory of natural selection. It was, for several decades, mostly a matter of different interpretations of the same observations. Of course, the better approach was to gather more evidence that would weigh for or against natural selection as a cause, and such evidence has unfolded as biologists have studied mimicry in greater depth.

One of Bates' key inferences was that the species being mimicked is not palatable to predators, and that the palatable species gains protection by imitating it. This would imply that predators learn or know to avoid the unpalatable form.

Figure 4.3 Caterpillar mimic of snake head. First discovered by Bates, a number of species mimic the appearance of snake heads. This is the Spicebush Swallowtail caterpillar (*Papilio trollus*). Photo by Mary Jo Fackler.

There have been many subsequent investigations of this suggestion, but the most notable controlled studies were conducted by Jane van Zandt Brower beginning in the late 1950s. Using wild caught birds, Brower showed that many species thought to be unpalatable were rejected and avoided by birds relative to mimetic and nonmimetic species. Furthermore, the birds showed a tendency to learn rapidly to recognize unpalatable butterflies and to reject those resembling them.

A second basic aspect of mimicry is the expectation that protection from predators should break down where the unpalatable form is not present. This prediction has recently been tested in a fascinating example of mimicry among snakes.

The harmless and very beautiful scarlet kingsnake and Sonoran mountain kingsnake resemble poisonous coral snakes in that each species possesses red, yellow, and black ring patterns. The sequence of color bands differs between the harmless and poisonous snakes according to the helpful rhyme repeated by many snake enthusiasts:

> Red touch yellow, kill a fellow.
> Red touch black, friend of Jack.

David and Karin Pfennig and William Harcombe of the University of North Carolina (Chapel Hill) identified dozens of sites in North Carolina, South Carolina, and Arizona where both coral snakes and kingsnakes occur together as well as where the coral snake does not occur. At each site, they left ten sets of three snake replicas made of plasticene (a soft, nontoxic clay): one with a tricolor ringed pattern, one with a striped pattern of identical colors, and one with a plain brown pattern. After weeks in the field, the plasticene replicas were collected and scored for predator bite and scratch marks by a scientist who did not know whether they had been placed in areas where the snakes overlapped or where the poisonous species was absent.

It turned out that at the Carolina sites, the proportion of kingsnake replicas attacked was much higher (68 percent) in areas where coral snakes do not occur than in areas where they do occur (8 percent). Similar results were found at the Arizona sites, confirming that predators avoid coral snake mimics in areas where coral snakes live.

As Darwin predicted, Bates' work has had lasting value. Biologists to this day refer to the imitation of unpalatable or poisonous forms by palatable, harmless species as Batesian mimicry.

Bates never returned to the Amazon, but he did finish the book of his travels, *The Naturalist on the River Amazons* (1863). When it was published, he sent a copy to Darwin and anxiously awaited the verdict from the author of *The Voyage of the Beagle*. Darwin replied, "My criticisms may be condensed into a single sentence, namely, that it is the best work of Natural History Travels ever published in England."

It is still a great read today—full of tales of adventure and great descriptions of the animal and human residents of the Amazon. When describing butterfly wings, which became to Bates what the Galápagos finches represented to Darwin, Bates wrote poetically, "It may be said, therefore, that on these expanded membranes nature writes, as on a tablet, the story of the modifications of species."

PART TWO

THE LOVELIEST
BONES

The Origin of Species *set a new agenda for paleontology. While fossils were becoming better known and studied in the preceding decades, their meaning was not understood. Indeed, one of the great ironies of the early history of paleontology is that dinosaur fossils were well known before the publication of* The Origin, *having been named and studied at length by the anatomist Richard Owen, but Owen viewed them as evidence* against *evolution. Before Darwin, most paleontologists had little sense of the great antiquity of the planet, of life, or of the bones they held—the deep history of nature.*

Darwin, thoroughly grounded in the emerging new geology, changed all that. In his book he suggested that the time span that separated species or geological strata was not a mere thousand generations, as previously thought, but a million or a hundred million generations. He wrote, "The crust of the earth is a vast museum," of which only a small portion had been explored. He asserted that the ancestors of living species were buried in the rocks of the earth, so finding these supposed ancestors was crucial.

With remarkable candor, Darwin admitted the lack of fossil evidence at the time for critical parts of his theory, such as the absence of transitional forms between major groups of animals or the sudden appearance of animals in the fossil record without any hint of the gradual development of the simpler forms preceding them. There was also the very delicate question, which Darwin dodged, of human antiquity. But he did not fool anyone, as human origins were on everyone's mind from the moment his revolutionary book appeared. Several of the boldest expeditions and greatest discoveries in paleontology have filled in the gaps and established the links between specific groups, such as fish and amphibians, reptiles and birds, and apes and humans.

I will begin with one of the most single-minded expeditions in all of paleontology—Eugène Dubois' search for ancient humans, which prompted him to abandon a medical career in the Netherlands for the malarial heat of Indonesia (Chapter 5). Inspired by Darwin's new theory, Dubois decided that the most important thing he could do was to find a "missing link" between apes and humans. His "Java Man" was the first claimed link and a portent of the heated controversy that would greet almost every subsequent hominid fossil and claim.

The second story centers on the apparently sudden emergence of animal life in the Cambrian fossil record that so worried Darwin. It was the search for progressively older fossils and the dawn of animal life that led Charles Walcott to two great discoveries (Chapter 6). First, in the depths of the Grand

Canyon, he discovered the first clear evidence of life before the Cambrian, which showed that life had originated much earlier and in simpler forms. And second, on top of the Canadian Rockies in the Burgess Shale, he found the greatest assortment of some of the earliest and most peculiar animals of all. These creatures document "the Cambrian Explosion"— the fairly rapid appearance of large, complex animals over 500 million years ago and the early origins of major divisions of the animal kingdom.

For sure, the most awe-inspiring animals in all of the fossil record are the dinosaurs. Perhaps the greatest natural history expedition ever mounted, the exploration of the Gobi desert of Mongolia, led by the intrepid Roy Chapman Andrews (Chapter 7), was not initially trying to find dinosaurs at all but hoped to find ancient humans. They wound up discovering the first dinosaur eggs, Oviraptor, Velociraptor, and other dinosaurs in great numbers, as well as the earliest fossil mammals then known but, alas, not a trace of ancient humans.

The pursuit of dinosaurs has provided much more than museum attractions; it has produced some startling scientific insights into the origin and extinction of major groups of animals. The disappearance of the dinosaurs at the end of the Cretaceous period was well known to early paleontologists.

But the cause remained a mystery for many decades until a remarkable father-and-son team, a physicist and a geologist, found the first clues in a thin layer of clay outside a small town in Italy. In Chapter 8, I will tell the story of the detective work that took scientists across the globe to find the cause of this great extinction—and one of the most important and revolutionary discoveries in geology, paleontology, and biology in the twentieth century.

While that extinction was the end of the great behemoths, as well as perhaps more than 80 percent of all other land species, it turns out that it was not the end of the dinosaurs as a group. New dinosaur fossils discovered in the 1960s, and a reexamination of key fossils discovered in the nineteenth century, inspired a renaissance in dinosaur research and a revolution in thinking that led to the realization that dinosaurs were not extinct and that birds are, in fact, a type of dinosaur (Chapter 9).

The search for "missing links" in the evolution of animal forms has not abated. Recent expeditions to unexplored regions of the globe have turned up some other startling creatures that reveal major evolutionary transitions. One of the most spectacular transitional fossils ever found was recently unearthed in the Arctic and reported in 2006. The "fishapod," an animal with characteristics of both fish and four-legged vertebrates, throws new light on one of the major events in all of animal history, the transition of vertebrates to life on land (Chapter 10).

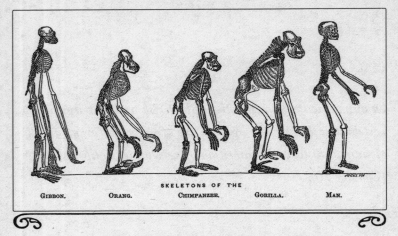

SKELETONS OF THE

GIBBON.　　ORANG.　　CHIMPANZEE.　　GORILLA.　　MAN.

Figure 5.1 The evolution of apes and humans. The famous frontispiece from T. H. Huxley's *Evidence as to Man's Place in Nature* (1863).

∾ 5 ∾
Java Man

No great discovery was ever made without a bold guess.
— *Isaac Newton*

HIS COLLEAGUES THOUGHT he had lost his mind.

What else could explain why this promising young physician and anatomist, who was sure to rise to a prestigious professorship at the leading medical school in Holland, threw it all away for a posting with the Dutch army 10,000 miles away in the East Indies? What's more, how could he in good conscience take his beautiful young wife and new baby to such a faraway, foreign, and dangerous place?

What did twenty-nine-year-old Dr. Marie Eugène François Thomas Dubois, born the very year that Wallace's and Darwin's first papers on natural selection were announced to the world, hope to find in the East Indies?

The most important discovery he imagined that anyone could make in the early days of evolutionary theory — the "missing link" between apes and Man. It would be the ultimate crown on the theory of evolution, the definitive proof of the link between humans and the rest of the animal world. And the name Eugène Dubois would be famous and respected all over the world.

The journey was, in his mind at least, not a flight of fancy but a matter of destiny. Dubois believed that everything he had learned, from his boyhood exploration of the plants and fossils in his native Limburg, through his early schooling and later medical training, to his current anatomical research on the human larynx, had prepared him for that great quest.

As a child, Dubois was absorbed in the study of nature. He made frequent trips into the countryside to collect medicinal herbs for his father's pharmacy. From his bedroom he could see St. Peter's mountain, near Maastricht, a rich fossil chalk bed famous for the discovery in 1780 of the first Mosasaur (from the Latin *"Mosa,"* meaning "the Meuse river" in the Netherlands, and the Greek *"sauros,"* meaning "lizard"), a late Cretaceous marine reptile. Dubois went on many expeditions to collect fossils in the vast chalk formations.

From his early school days, the question of human origins swirled about him. When he was just ten years old, Dubois heard about a series of lectures by the zoologist Carl Vogt that stirred up controversy throughout Holland. A staunch supporter of Darwin's new theory, Vogt embraced the view of humans as being members of, not above, the animal kingdom. In high school, Dubois' science master introduced him to the great works that developed and promoted these ideas—by Darwin, Thomas Huxley, and Ernst Haeckel.

Darwin and Alfred Wallace had solved "the mystery of mysteries." Dubois focused on what Huxley described in *Evidence as to Man's Place in Nature* (1863) (Figure 5.1): "The question of questions for mankind—the problem which underlies all others, and is more deeply interesting than any other—is the ascertainment of the place which Man occupies in nature and of his relations to the universe of things."

This book was the first detailed biological examination of humans in the light of the emerging comparative studies of apes and the first fossil humans, which Huxley had pursued intensely. While Darwin had stated in his work, "Light will be thrown on the origin of man and his history," he deliberately avoided any further discussion of human origins, figuring (correctly) that his great theory had many objections to overcome without raising the delicate issue of human evolution.

But Huxley picked up where Darwin left off and confronted the central issues head-on. He pleaded for a dispassionate, objective, zoological approach and asked his readers to approach human biology as someone from another planet would:

Let us . . . disconnect our thinking selves from the mask of humanity; let us imagine ourselves scientific Saturnians, if you will, fairly acquainted with such animals as now inhabit the Earth, and employed in discussing the relations they bear to a new and singular "erect and featherless biped," which some enterprising traveler, overcoming the difficulties of space and gravitation, has brought from that distant planet for our inspection, well preserved, may be, in a cask of rum.

He then asked:

. . . is man so different from any of these Apes that he must form an order by himself? Or does he differ less from them than they differ from one another, and hence must take his place in the same order with them?

Huxley urged his readers:

Being happily free from all real, or imaginary, personal interest in the results of the inquiry thus set afoot, we should proceed to weigh the arguments on one side and on the other, with as much judicial calmness as if the question related to a new Opossum.

Dubois saw that such "judicial calmness" was in short supply.

Huxley's analysis of human bodies, brains, and eggs built the zoological case, but he also brought new paleontological evidence into the argument. The then newly discovered Neanderthal remains and a second fragmentary skull found in Belgium along with the remains of mammoths and a woolly rhinoceros were, to Huxley, definitive evidence of human antiquity.

Others were skeptical, if not outright hostile, to the idea of ancient humans. The leading German pathologist, Rudolf Virchow, concluded that the Neanderthal's unique skeletal features were deformities caused by disease, and the remains were not those of a distinct human race or species.

Dubois' interest in "the question of questions" was further stoked by the ideas of the German embryologist Ernst Haeckel. In his *History of Creation* (1868), Haeckel outlined a speculative history of human origins, beginning with a simple, single-celled ancestor. Building on Huxley, Haeckel underscored what he thought were the two most important adaptations that made humans distinct—walking upright and articulate speech. These abilities were made possible, Haeckel asserted, by two major morphological changes in *"the two pairs of limbs and the differentiation of the larynx."* Haeckel proposed that walking upright long preceded the acquisition of speech and that there was a stage in the evolution of human ancestors that he called "Speechless Man (Alalus), or Ape-Man (Pithecanthropus), whose body was indeed formed exactly like that of a Man in all essential characteristics, but who did not as yet possess articulate speech."

Neither Haeckel nor Huxley viewed the Neanderthal as an intermediate between "Men and Apes," and Huxley closed his book asking:

Where, then, must we look for primaevel Man? . . . In older strata do the fossilized bones of an Ape more anthropoid [like a human], or a Man more pithecoid [apelike], than any yet known await the researches of some unborn paleontologist?

Time will show.

That paleontologist was not unborn at the time Huxley wrote those words, but he was just five years old. And later, as he read those words over and over again, his determination to be that paleontologist grew.

Amsterdam

Dubois' father wanted him to follow in the family business and become a pharmacist. But Eugène, a fiercely independent young man, was determined to continue his studies of nature. That meant going to medical school, where the first year of studies was focused entirely on the natural sciences. In 1877, at the age of nineteen, Dubois enrolled at the University of Amsterdam.

The outstanding faculty included luminaries such as the physicist Van der Waals (Nobel Prize winner in Physics, 1910), the chemist Van't Hoff (winner of the first Nobel Prize in Chemistry), and the botanist Hugo de Vries (one of the rediscoverers of Mendel's work on heredity). De Vries and Dubois often discussed the raging debate over human origins.

Dubois soon realized that he had little interest in becoming a practicing physician, but he did not take Darwin's way out. He worked hard, was disciplined and focused, and did very well at every subject

Figure 5.2 **Eugène Dubois, age twenty-five.** Photo courtesy of and copyright the National Museum of Natural History, Leiden, Netherlands.

he touched. His talent was recognized, and in 1881 Dubois was of-fered an assistantship in anatomy by Dr. Max Fürbringer. It was a stroke of luck, as Fürbringer had himself been trained by Haeckel. Fürbringer helped Dubois gain promotions—from assistant, to pro-sector in charge of the anatomy course, and then to lecturer—in rapid succession. It was a meteoric rise: Dubois was just one rank below that of full professor at the age of twenty-eight.

Dubois decided to start an independent research program on the comparative anatomy of the larynx, the structure responsible for the uniquely human capability of articulate speech. He published one paper, but a confluence of forces soon turned him away from his academic career and toward the East Indies.

The first was his discovery that he hated teaching. He was so anxious before lectures that he talked to no one and wanted none of his colleagues to attend them. The second was a falling-out with Fürbringer. Dubois was very ambitious and eager to gain recognition for his work. When he gave a draft of his first larynx study to Fürbringer, his mentor commented that he had previously made some of the same points. Dubois worried that his work would not be seen as his own. He revised the article but never stopped stewing over it, and he grew increasingly suspicious of Fürbringer's motives.

The third force was new fossil finds, which reignited Dubois' interest in human paleontology. In 1886, near Spy, Belgium, more Neanderthal remains were found. There was no doubt that these were old bones, and they demolished the argument that the original find in Germany had been that of a diseased individual. No, the Neanderthal and Spy fossils were something different from modern humans, but still a long way from apes or an apelike ancestor.

Dubois worried that the years were flying by and that others must be getting close to finding the missing link. If he was to seize the prize, he had to act.

Where to?

Dubois made up his mind to chuck it all—the certainty of a professorship, the teaching, the anatomy lab, Fürbringer—to look for the missing link. The trouble was—where to look?

Certainly not in Europe, where the Neanderthals had been found. Darwin had suggested in *The Descent of Man* that since humans lost their fur covering, they must have originated in the Tropics, not colder zones. That excluded North America but left Africa, Asia, part of Australia, and South America. But, since apes were found only in tropical parts of the Old World, it followed that human ancestors originated in those same regions. That left Africa or Asia.

Dubois knew that Darwin favored Africa because of humans' affinities to gorillas and chimpanzees. But Asia held the gibbon and orangutan, and Haeckel had argued that gibbons were more closely related to humans. Furthermore, a fossil ape known as the Siwalik chimpanzee had recently been discovered in the Siwalik hills of British India.

The age and location of this find suggested that deposits of a similar age might be fruitful, and studies by a Dutch paleontologist indicated that such deposits might be found in Borneo, Sumatra, or Java. Dubois was well aware of Wallace's "line" and Wallace's work on the geographic distribution of animals. He knew that the animals of the western part of the Malay Archipelago were shared with those of mainland Asia, so whatever was found in India could also occur in these islands.

Furthermore, all the human fossils that had been found to date were found in caves, and Sumatra was littered with caves. And one very practical factor pushed him to Sumatra: It was part of the Dutch East Indies. He would find fellow countrymen and some familiar customs—he might even get some government support for an expedition.

Dubois made a pitch to the secretary-general of the Colonial Office. He laid out his logic behind the expedition and the glory the discovery of the missing link would bring to Dutch science. But the secretary-general explained that there was no money for such a speculative venture.

Dubois had to support himself and his family, but what was he able to do? He did have a skill that was needed in the Dutch colonies, for he was a physician. The Dutch army needed him, so he enlisted—*an eight-year commitment.*

When he told his wife, Anna, she was surprisingly supportive. But his family and in-laws were not. His father thought he was throwing away a distinguished career.

But there was no stopping him. He and his family would sail to Sumatra (see Figure 3.2 for map).

Sumatra

The voyage from Amsterdam to Padang took forty-three days, even with a shortcut through the Suez Canal. Dubois and his family arrived on December 11, 1887, and tried to settle in among the exotic sights and scents of Sumatran life. It was the rainy season, and Eugène and Anna, pregnant with their second child, quickly learned what that meant—buckets of rain every day and mud everywhere. Anna tended to the setting up of the household, while Eugène reported to the army hospital.

The conditions there were unlike anything Dubois had seen or imagined in Holland. He was overwhelmed by patients suffering all sorts of fevers: from cholera, malaria, typhus, tuberculosis, and other unidentifiable maladies. The amount of work was so great that he was not sure when, or if, he would get into the field to prospect for bones.

He made do for a while just telling his fellow officers about his reason for being there—the missing link—and put together a lecture explaining his logic. He then used it as an outline for an article that he wrote for the *Journal of the Natural History of the Netherlands Indies.* The article both staked his claim to his search and served as a warning to the government authorities to support this scientific work or see the glory reaped by another nation.

His forays into the countryside near the base were not promising, so he requested a transfer to another, more remote hospital, with more caves to look into and fewer patients to look after. Anna, now eight months pregnant, had to start all over in putting together a new home, but the move to the cooler highlands was a relief from the swelter of Padang. She gave birth to their first son at home in Pajakambo, attended to by Eugène.

With more time to search, Dubois had some luck. He found a cave called Lida Adjer that had a good many bones of rhinos, pigs, deer, porcupines, and other Pleistocene animals. In the meantime, his article drew the attention of the governor, who promised him some laborers for his explorations. Dubois wrote to the governor about his new finds. Even several colleagues back home were behind his work and encouraging the government to support it.

In March 1889, the Dutch government authorized the assignment of two engineers and fifty laborers to help with the search and excavations of the large number of Sumatran caves. At last, Dubois thought, he could work properly. Surely, finding the missing link was just a matter of time.

But many of the caves were empty or inhabited by animals that were far from petrified. One day, Dubois was frustrated at the workers' reluctance to enter a cave, so he crawled into its narrow passageway. Penetrating farther, he was overwhelmed by the smell of cat urine and rotting meat—it was a tiger's lair. Dubois tried to back out quickly but became stuck and had to plead with his workers to pull him out.

Dubois shook off the episode, but he couldn't shake off other dangers in this land. He was felled by a bout of malaria, the first of many that would interrupt his work (if they didn't kill him outright). Malaria soon felled many workmen. One of the two engineers died of the fever, and half of his workmen were too sick to continue. Others quit and ran off. Months passed without success, and Dubois, having spent two years in Sumatra, wrote to the director of the National Museum of Natural History in Leiden:

> Everything here has gone against me, and even with the utmost effort on my part, I have not achieved a hundredth part of what I had visualized . . . I have found a few very useful caves, but still never the best one could wish for. What's more, it was necessary to live out in the forest for weeks on end, usually under an overhanging rock or in an improvised hut, and it turns out that in the long run I can't stand up to that, however well I was able to bear the fatigue at first. Having now come back, with my third bout of fever, which nearly finished me . . . I have had to give it up for good.

Dubois began to rethink his plans. He had heard from a geologist that the fossils on the island of Java could be older than those he was finding on Sumatra. Moreover, a petrified human skull had been found there the previous year in a rock shelter, a sign that Java might be more productive. Dubois put in for and received a transfer to Java.

Java Man

The family, four in all, packed up and went to Java. They found a good house in the town of Toeloeng Agoeng and settled in. Far from any army base, Dubois could follow his scientific pursuits full-time. He was assigned a new crew, led by two corporals who were far more competent than his engineers on Sumatra.

He started his excavations in June 1890 at Wadjak, where the human fossil skull had been found two years earlier, and was quickly rewarded. His crew found all sorts of extinct mammals, including rhinos, pigs, monkeys, antelopes, and even another partial human skull.

Dubois then made a decision that proved critical. He expanded the

Figure 5.3 Excavation at Trinil on Solo River. Photo circa 1900, courtesy of and copyright the National Museum of Natural History, Leiden, Netherlands.

search beyond the caves and rock shelters in the hills to the riverbanks, for in the dry season, when the waters were low, the sediments of the riverbanks were exposed (Figure 5.3). In the hills and along the banks of the Solo River, Dubois' team came across unusually rich deposits that included more rhinos and pigs, hippopotami, two different types of elephant, big cats, hyenas, crocodiles, and turtles.

Then, on November 24, 1890, they found a fragment of a human jaw with two teeth. Its decrepit state made further identification difficult, but it was an encouraging sign for the next year's work.

The bones were piling up on the veranda at home (Figure 5.4). Dubois wanted to write descriptions of the fossils and the places they were found, but the magnitude of the task was overwhelming and still growing. And he did not yet have his prize, if it was to be found in Java, or anywhere else for that matter.

During the second season on Java, digging began at Trinil, on the Solo River. In September 1891, workers unearthed the third molar tooth of some primate. Dubois figured it for that of a chimpanzee like the Siwalik hill type.

The Trinil site was rich and produced many more mammal fossils. The next month, the engineers unearthed a bone they took for part of a turtle shell. Concave and a deep brown color, the fossil was sent to Dubois at his home.

There was no doubt this was not a shell piece but part of a skull, a primate skull. It had a brow-ridge like that of a chimpanzee (Figure 5.5). but it encased a larger brain than a chimp skull would. Dubois decided that it was from some kind of ape. He would need other skulls to make some detailed comparisons, so he requested a chimpanzee skull from Europe.

Dubois was anxious to find more fossils, but the season drew to a close. He spent the winter cleaning the skullcap and securing other skulls so he could make some detailed comparisons with his new find.

By May 1892, excavation work had resumed at Trinil. First, the silt from the long rainy season had to be removed from the plot that Dubois and his engineers had staked out. Dubois spent more time at the site, but by the end of July, after a two-week stint, he was worn

Figure 5.4 The fossils piling up on Dubois' veranda. Photo courtesy of and copyright the National Museum of Natural History, Leiden, Netherlands.

out. He wrote in his diary: "I discover that there is no more unsuitable place available in Java, because of health and malaria, for the study of fossils than this hell." He then developed another bout of fever.

The next month, the engineers again made a remarkable find—this time an almost complete left thigh bone. When the piece reached Dubois, he was delighted. He could tell that this animal was in no way equipped to climb trees. It was very like a human (Figure 5.6).

Now he had a molar, a skullcap, and a thigh bone. It made perfect sense to Dubois that these fossils, found relatively near one another though at separate times, came from one individual. The thigh bone was crucial. Its features indicated that it came from an upright walking ape and therefore was a new species. He named his find *Anthropopithecus erectus* Eug. Dubois—the upright walking chimpanzee.

But he soon discovered he had made a mistake, a wonderful mistake. When he first estimated the volume of the braincase from the skullcap, he obtained a figure of 700 cc: larger than a chimpanzee's (410 cc), but much smaller than a human's (c. 1,250 cc). But he had

Figure 5.5 The skullcap found at Trinil. Photo courtesy of and copyright the National Museum of Natural History, Leiden, Netherlands.

Figure 5.6 **The thigh bone found at Trinil.** Photo courtesy of and copyright the National Museum of Natural History, Leiden, Netherlands.

measured the skullcap and figured the braincase capacity incorrectly. A recalculation put the skullcap closer to 1,000 cc, much larger than that of any ape and much closer to, but not fully, that of a modern human. His fossil was not an ape, not a human, but an upright-walking *intermediate* between apes and humans.

He had done it.

Five years after arriving in the East Indies, having left his job, parents, and Holland and after scouring countless caves, dodging tigers, and battling bout after bout of malaria, he had found the missing link.

He renamed his find *Pithecanthropus erectus*, "erect ape-man." It was time to tell the world.

The world reacts

If this were a Hollywood movie (and maybe it ought to be!), it would end here, and we could all walk out of the theater smiling, having seen that Dubois' bold gamble, damaged health, hard work, and his family's many sacrifices were rewarded with great luck and the prize he had sought. And we could be certain that fame and the acclaim of the scientific world were sure to follow.

But that did not happen. To claim such a prize as the missing link, Dubois and his fossils had to withstand a storm of critical scrutiny;

some of it was good, honest scientific analysis, as it should be, and some not.

For most of 1893, Dubois worked on putting together a description of *Pithecanthropus*. He first thought he could write a series of articles on his work in Java, with the ape-man as part of it. But that would require dealing with the tens of thousands of fossils he had collected. He soon decided that his prize specimen must get his full attention.

Dubois' thirty-nine-page description included photographs of the thigh bone and skullcap and comparative illustrations of other ape skulls. Dubois emphasized the close proximity of the places where the molar, skull, and thigh bone had been found and expressed his strong conclusion that the remains were all from the same individual. Examining details of the skull, he pointed out both its several human and apelike features and its large capacity.

On the basis of the thigh bone, he argued that *Pithecanthropus* walked the way humans do and was about the same height and body size. Altogether, his ape-man was just that—something in between ape and man. He stated: "*Pithecanthropus erectus* is the transitional form which, according to the theory of evolution, must have existed between Man and the anthropoids; he is Man's ancestor."

Dubois had the work printed in Batavia, the capital of the Dutch East Indies, and it reached Europe by the end of 1894.

He did not have to wait very long for reactions from Europe, but they were not what he expected or hoped for. Criticism poured in from many directions. A German anatomist declared the skullcap undoubtedly that of an ape and the femur, human. He credited Dubois only with finding a fossil gibbon and more evidence of the antiquity of humans. Rudolf Virchow, the outspoken skeptic on Neanderthals, also concluded that the skullcap came from a gibbon and refused to accept *Pithecanthropus* as a missing link.

Other commentators took a different angle. Some British scientists saw the skull as human. The paleontologist who described the Siwalik chimpanzee wrote a critique in the prominent journal *Nature* that suggested the skull was that of a diseased microcephalic human (a condition that arrests brain and skull growth). Some suggested that the bones could have come from a member of a primitive human race and were not a transitional form.

There were some exceptions. The American paleontologist O. C.

Marsh thought that Dubois had proven his case. And Ernst Haeckel was, not surprisingly, supportive.

But the majority of opinions were negative. They stung. Here Dubois was halfway around the world, living and working in primitive conditions, making actual discoveries, while these European academics were sitting in the comfort of their lofty offices. They had some gall writing papers and giving lectures on fossils they had never seen and doubting the analysis of the one man who had found and studied them! Dubois concluded that the ferocious criticism must be caused by jealousy. He had found the missing link, and now they wanted to deny him his rightful credit!

He had to return to Europe and convince the doubters.

Home

The family had nearly eight years' worth of possessions to sort through to decide what to ship home, what to carry with them, and what they would leave behind in Java. Dubois had 414 crates containing more than 20,000 fossils, but there was only one possession on his mind, his *Pithecanthropus*. He had a special wooden suitcase made to hold the two wooden boxes that cradled the precious fossils, and he carried it himself to Batavia, on the ship to Marseilles, and on the long train to Amsterdam.

During the six-week journey, Dubois prepared, both strategically and psychologically, for the battle ahead. But Nature decided he must endure one more challenge. Out on the Indian Ocean, the ship was engulfed in a terrific storm. The captain ordered the passengers into the lifeboats in case they had to abandon ship.

Just then, Dubois remembered that he had left *Pithecanthropus* in his cabin. If he were to lose it, he would have nothing to show for all of his efforts and nothing with which to defend himself against the backlash brewing in Europe. He went back to get it and told Anna that if the boats were lowered, she was to look after the children. He would hold on to his newest child—*Pithecanthropus*.

Fortunately, the storm passed, and no such separation was necessary. By early August, the family reached Holland.

For Dubois, it was not a triumphant reunion. His father had passed away while he was in Java, so he did not have the satisfaction of proving that his long journey was worthwhile. But neither was his mother impressed by his box of bones. Seeing *Pithecanthropus*, she asked, "But, boy, what use is it?" Some of his critics were asking the same question.

Dubois hurled himself into a campaign to persuade all of Europe of the importance of his discovery and the correctness of his interpretation. He and *Pithecanthropus* commenced a tour of scientific conferences and the most important institutions in Germany, France, Belgium, and Great Britain.

His first opportunity came in just a few weeks, at the International Congress of Zoology being held in Leiden. Many important figures were present, including his arch-critic Virchow, who presided. Dubois knew he must be in top form to stand up to Virchow's unwavering skepticism and outright ridicule. Earlier that year, Virchow had described Dubois' interpretation of the Java fossils as a "fantasy . . . beyond all experience."

Dubois wisely avoided any personal attacks and focused instead on the legitimate scientific questions that had been raised about the fossils. He acknowledged that his thirty-nine-page description was inadequate on some important matters and attempted to fill in the gaps. He described more fully the geological formation in which the fossils had been found and the other animal bones in the same layers. He emphasized the human characteristics of the thigh bone, the apelike features of the skullcap, and its intermediate braincase capacity.

Virchow was unmoved, but others acknowledged that Dubois was clearing up some of the misconceptions and unanswered questions from his original report.

Most important, Dubois allowed many potentially influential scientists to inspect the bones for themselves. That warmed a few to his side. In Paris, he developed a great ally, Professor Manouvrier. He traveled to Edinburgh and Dublin and won over a few more converts, but by no means did he yet have a favorable consensus.

Dubois was buoyed by some official acclaim: The Anthropological Institute of Great Britain and Ireland made him an honorary fellow, and in 1896 he received the Prix Broca in Paris for his achievement.

While all of his traveling had been necessary, Dubois wanted to settle down and establish a base from which he could continue his work. By an act of parliament, he was appointed curator of paleontology at the Teyler Museum in Haarlem. His family moved yet again, and he kept writing papers about and campaigning for his ape-man.

In 1898 the most prominent luminaries of biology—including supporters, critics, and the undecided—gathered at the International Congress of Zoology in Cambridge, England. Ernst Haeckel, now an elder statesman of evolutionary science, took the podium before Dubois and used all of his considerable eloquence to support *Pithecanthropus* and to vanquish the old guard of Virchow and his fellow hold-outs. Pronouncing Dubois the "able discoverer of *Pithecanthropus*" who had "convincingly pointed out his high significance as a missing link," Haeckel made such a formidable case that Dubois just sat back and enjoyed the show.

Haeckel laid into Virchow in particular, enumerating point by point how experts disagreed with him and underscoring how Virchow had declared Neanderthals and *Pithecanthropus* to be "pathological products; indeed the sagacious pathologist at last made the incredible assertion that 'all organic variations are pathological.'" Haeckel continued that "it must be remembered that for more than thirty years Virchow has regarded it as his especial duty as a scientist to oppose Darwinian theory [and to assert that] it is quite certain that man did not descend from the apes . . . not caring in the least that now almost all experts of good judgment hold the opposite conviction."

As gratifying as Haeckel's support was, the campaign was not over, and Dubois kept searching for ways to convert others to his view of *Pithecanthropus*. He developed a new approach to the argument by comparing relative brain and body sizes in animals. He found a general mathematical relationship between the two in most mammals. From these data, he asked, "What size ape would have a brain almost 1,000 cc in volume?" From his studies, he calculated that it would have been about a 500-pound ape. But from the dimensions of the *Pithecanthropus* femur, he could tell that it would support a roughly 160-pound animal. The brain was too big to be that of an ape, yet it was smaller than that of a modern 160-pound human. It was intermediate in size, which is just what one would expect if, as he had now been arguing for nearly five years, *Pithecanthropus* was indeed an intermediate between apes and humans.

Homo erectus

It had been over a decade since Dubois, Anna, and baby Marie had boarded the ship for the East Indies. The years of criticism and debate, following the years spent searching all over Sumatra and Java, had taken a toll on Dubois. He was battle weary and physically worn down. Moreover, he felt he had said about all he could about the fossils in hand; they no longer held his fascination the way they once did. As the century drew to a close, he had to satisfy himself with the knowledge that he had convinced many, but not all, of the place of *Pithecanthropus* in human history and of his own claim to a place in paleoanthropological history.

But even Dubois did not yet know the full magnitude of his feat. Many followed in his footsteps to Asia over the next several decades, looking for ancient humans. Their searches demonstrated how unlikely it was that Dubois had found anything at all, as most found no hominid fossils whatsoever. Further Dutch and Prussian excavations at Trinil turned up nothing. Later, the largest land-based expedition ever mounted, a decade-long search for ancient humans in China and Mongolia led by the American Museum of Natural History, found no human fossils (but was a smashing success in unexpected ways; see Chapter 7).

It was almost forty years before anyone found more evidence of ancient hominids in Asia. In 1929–30, "Peking man" (given the name *Sinanthropus pekinensis*) was described from caves in China, and in the late 1930s more *Pithecanthropus* skulls were finally found on Java. In 1950, these two fossil hominids were joined together as one species, placed in the same genus as humans, and renamed *Homo erectus*. Dubois' skullcap, known as Trinil 2, is now the type specimen of *Homo erectus*—the original specimen on which the description of a new species is established.

Dubois also could not have known that his battle was typical of the contentious reaction that would greet the identification of virtually every new hominid fossil and the attempts to place them in the history of human origins. Another impact of his discovery was that most anthropologists were convinced, because of the antiquity of *Pithecanthropus* and a host of other reasons, that the cradle of mankind was in Asia, not in Africa, as Darwin had surmised. The discovery of ancient hominids in South Africa in the 1920s was therefore dismissed and ignored. It was not until the early 1960s, when older *Homo erectus* and other fossil hominids were found in Africa, that the focus of human origins shifted fully away from Asia to Africa (see Chapter 11), and new methods, beyond the wildest imaginations of the great fossil hunters of the past century, would be brought to bear on the "question of questions" (see Chapters 12 and 13).

Figure 6.1 A meeting of legends. February 10, 1910, on the street in front of the Smithsonian Institution in Washington, D.C., Charles Walcott (left) escorts (from left to right) Wilbur Wright, Alexander Graham Bell, and Orville Wright to a waiting car. Photo courtesy of the Smithsonian Institution Archives (SIA 82–3350).

∽ 6 ∽

To the Big Bang, on Horseback

We are what we repeatedly do.
Excellence, then, is not an act, but a habit.
— *Aristotle (384–322 B.C.)*

ON THE AFTERNOON OF February 10, 1910, in Washington, D.C., three men in top hats and their finest attire strode to the curb to board a sleek new horseless carriage. They paused for a moment that was captured by a photographer, three inventors whose deeds would become legendary and whose names would be synonymous with their inventions—Wilbur and Orville Wright and Alexander Graham Bell (Figure 6.1).

Escorting these illustrious figures to their car was Charles Doolittle Walcott, a man whose deeds and name are much less known to posterity, if at all. One might think that rubbing elbows that day with such famous men would have been the highlight of Walcott's year, perhaps of his whole life.

It was not even a close second.

Six months earlier, while riding his horse high in the Canadian Rockies over Burgess Pass, this nearly sixty-year-old veteran geologist unearthed one of the oldest and most important mother lodes of animal fossils ever found. The remarkable and odd animals of the Burgess Shale mark one of the greatest, and to Darwin one of the most troubling, mysteries in paleontology—the Cambrian Explosion, the apparently sudden emergence of large, complex animals in the fossil record more than 500 million years ago.

There was another reason that Walcott was so nonchalant about his company that day. He was quite used to circulating among the country's elite. Though he never finished high school or earned any kind of diploma or degree, he rose to be the director of the U.S. Geological Survey, secretary of the Smithsonian Institution, a founder of the Carnegie Institution of Washington, and president of the National Academy of Sciences. By February 1910, he had already known and advised four presidents and would come to serve the next three. It was Walcott who helped to persuade President McKinley to set aside national forest reserves and who conspired with Teddy Roosevelt to designate lands for national monuments and parks. And he carried out all of those duties while still managing to explore much of the geology of North America and to make, not one, but two landmark discoveries in paleontology.

His remarkable story and ascent began with exactly what marked his crowning discovery on that Canadian mountain peak: trilobites.

Growing up in the Cambrian

Born in 1850, Charles Walcott grew up in Utica, New York. Having lost his father when he was two and become an adolescent in the middle of the Civil War, he grew up quickly. With many men away serving in the Union forces, there was a lot of work available for the young Walcott.

He started out as a summer hand on William Rust's farm near Trenton Falls. Dairying was Rust's main source of income, but like other farmers in the area, he maintained a small limestone quarry that supplied house and barn foundations as well as some extra income. The area was known for its Trenton limestone, which was rich in fossils, in addition to being an abundant building material. Though smaller than Niagara Falls, the Trenton Falls were also a tourist destination for New Yorkers, and the fossils sparked something of a collecting craze in the first half of the nineteenth century.

Walcott caught the bug early on. While quarrying was hard work, it did provide the dividend of many a fossil. He soon learned where in the limestone layers the fossils were most abundant, especially the best trilobites, which were more prized than the shelly remains of brachiopods.

By the age of seventeen, he was amassing a collection, and the long upstate New York winters gave him the opportunity to bone up on geology and paleontology. He learned how different fossils characterized different layers of rock and marked different geological eras. One day he found a block of sandstone on the road between Trenton and Trenton Falls. He broke out all of the fossils, but none of them were the Trenton limestone type he knew so well. By searching for the sandstone layer in the rock bed, he figured out that these trilobites must have been of pre-Trenton age, older than those he had been collecting.

Walcott deduced that they must have been from the Cambrian age —the earliest geological period then known to contain fossils. The name, from the Latin for Wales ("Cambria"), was conceived by the Reverend Adam Sedgwick after a series of field studies he conducted there in the 1830s, including his 1831 summer expedition to North Wales with his young assistant, Charles Darwin (see Chapter 2).

Walcott resolved at that moment that he would pursue the study of those older Cambrian rocks, but not in a formal way; Walcott found work, wages, and collecting fossils more appealing than school. He had another motive for sticking around Rust's farm—he had his eye on Rust's daughter Lura. By the spring of 1870, when Walcott was twenty, he worked on the Rust farm full-time and was engaged within a year.

Walcott went "trilobiting" as often as he could, scouring the Trenton Falls and Utica areas and traipsing about New York for new specimens, and started swapping fossils with other collectors near and far. He also kept up his reading, including Huxley's *Man's Place in Nature* and Darwin's *Descent of Man*.

In between nursing sick cows, painting barns, baling hay, and setting up a house with his new bride, Walcott was confident enough in his fossils and geology to contact some of the leading figures in academic paleontology and zoology. He was learning that there was a market for his collections, and he needed the money. The State Museum of New York, the Yale paleontologist O. C. Marsh, and the leading American zoologist, Louis Agassiz at Harvard, each expressed interest in his fossils. Walcott let Agassiz know that his collection had been deemed "the finest known from the Trenton group," and that it included 325 complete trilobites representing numerous genera, as well as crinoids, brachiopods, starfish, and corals (Figure 6.2). His asking price was $3,500 firm—the equivalent of several years' wages.

Figure 6.2 Trilobite. These trilobites (*Ceraurus pleurexanthemus*) were collected from the Trenton limestone by Walcott and sold to Louis Agassiz and the Harvard Museum of Natural History. Photo courtesy of the Museum of Comparative Zoology of Harvard University.

Agassiz did not bicker, and Walcott hauled the collection to Cambridge. Agassiz was a gracious host and received the young amateur as a fellow naturalist. When Agassiz died just a few months later, Walcott wrote to his widow:

> I looked to Professor Agassiz as a guide in whom I could trust and follow. I never knew what it was to have a father—most of my friends were bitterly opposed to my geological tastes. I had always fought my way and when Prof. Agassiz received and treated me as one doing what was right, he instilled into me an enthusiasm and determination to follow natural history as a pursuit that can never be eradicated.
>
> I am as yet but a boy of 23 years and shall devote the remainder of my life to this one object.

Walcott resumed collecting and filled his diaries and notebooks with drawings and descriptions of trilobites. In 1875, he published his first paper in a scientific journal, the *Cincinnati Quarterly Journal of Science*, describing a new species of trilobite. He followed it with a series of other short reports, each growing a bit in technical detail and sophistication. But these first successes were overshadowed by Lura's decline and eventual death in 1876.

Grief-stricken, Walcott was at a loss until James Hall, the chief geologist of the State Museum at Albany and a leading paleontologist, offered him a job as an assistant. Walcott left his tidy trilobite farm in Trenton Falls and moved on to much bigger adventures.

Down a grand staircase

It was a relatively brief apprenticeship in Albany, just a bit less than three years, but it was a valuable springboard. Hall was a difficult person, temperamental and very driven, and expected a lot from his subordinates. He had no reason to worry about Walcott, who threw himself into the exhaustive and systematic work; it was good therapy for his grief. Walcott learned more than paleontology. Hall was very well connected to politicians in the state legislature just down the street, and the museum depended on their goodwill and support. Walcott made sure his boss looked good whenever an important visitor came by to inspect the museum, a knack that served him well years later. Hall reciprocated by recommending Walcott for a position with the fledgling U.S. Geological Survey (USGS).

On July 21, 1879, Walcott, then twenty-nine years old, was appointed a geological assistant, just the twentieth employee of the USGS. He was assigned to a group who hoped to map out the little-known geology of the Grand Canyon and its surrounding regions. It had been just ten years since the one-armed Major John Wesley Powell and his companions had braved the rapids of the Colorado River and run the length of the Grand Canyon.

It was a five-day train trip to Utah, where Walcott then boarded a freight car on the Utah Southern Railroad and finally rode by stagecoach 120 miles to Beaver, Utah. He described the last stage as "the most tedious disagreeable ride I ever experienced."

His specific assignment was to map out the geology of a long, nearly unbroken series of geological formations that ran from the Colorado River in the Grand Canyon up to the so-called Pink Cliffs in southern Utah, a rise of some 8,000 feet. The cliffs, slopes, and terraces from the rim to the highest peak formed what Walcott's new boss, Captain Clarence Dutton, dubbed a "Grand Staircase," with each "riser" being a cliff or slope rising up to 2,000 feet and separated by "treads"—plateaus up to 15 miles wide. Walcott was to measure his way down the staircase, starting from the summit of the Pink Cliffs and eventually reaching the river. He would then identify each geological layer by the fossils it contained.

Walcott set out from Beaver by mule, traveling 10 to 15 miles a day. He was soon on his own, with just four animals and a cook. The USGS counted on the initiative of its geologists to figure out how to get their work done. Walcott was up to the task, though hard-pressed for time. The scenery certainly helped:

> The view from the summit of the White Cliffs is very fine but from the Pink Cliffs, nearly 10,000 feet above the sea, the scene is one to be long remembered. To the south the great basin of the Colorado lays before you. The White and Vermillion Cliffs form great escarpments carved in a hundred forms by the numerous canons. . . . The Pink Cliff terrace rises from the plain above the White Cliffs appearing in the morning sunlight as tho' a line of furnaces were in full blast as the pink rock . . . is lighted by the strong light.

Two months into his journey, he admitted: "Thus far I have enjoyed my trip very much. It is accompanied by many privations but still it will be a schooling for me in many ways."

A schooling for Walcott? It was more like a schooling for his fellow geologists. In a span of three months with just a hand level, a length of chain, and an altimeter, Walcott measured an 80-mile-long section more than 13,000 feet thick—a prodigious, unprecedented feat. Starting in the Eocene rocks of the Pink Cliffs (now the Bryce Canyon area), he found fossils from every major geologic period as he descended—the Cretaceous, Jurassic, Triassic, Permian, Devonian, and, at the Colorado River, a few trilobites that marked the Cambrian. As the weather turned snowy, he wrapped up his work, loaded up the mules, and carted his haul of more than 2,500 fossils back to civilization. It was a six-day trek to reach a railway.

His superiors were more than a little impressed. They doubled his salary and soon sent him west again. Walcott went to Nevada and the Eureka mining district, where he concentrated on paleontology.

Then, in the late summer of 1882, the new director of the agency, John Wesley Powell himself, beckoned Walcott back to the Grand Canyon.

Because of his 1879 survey, the top of the canyon was fairly well described, but its depths were not. Powell, a solid geologist, wanted to know more about the magnificent formations he had seen as he rushed through the canyon years earlier in his wooden boat. He decided the best way was to build a trail from the canyon rim down to the Colorado River. That way, Walcott and the others would be able to work during the winter months in the inner canyon, where it was warmer than up on the rim. That was the plan, at least.

Walcott and a small party headed to Kanab, Utah, from Nevada, an eight-day journey by train and wagon team, then slowly made their way south to the canyon. He was told to meet Major Powell at House Rock Spring, but he had become quite sick with what was probably a case of diphtheria. Walcott was trying to rest in a cabin when he spotted a bottle of turpentine and decided that it might help clear his congested throat. He took a gulp, burning his mouth and throat raw, but it did the trick. Though he was still too weak to help build the trail, he recovered in time to descend into the canyon and start what would be seventy-two straight days of geology.

With an assistant fossil collector, a cook, and a mule-packer under his supervision, Walcott began with a detailed study of the Tonto Group, a rock formation of Cambrian age. He then worked his way down, and down, and still farther down the geological sequence (Figure 6.3). Navigating the canyon was tricky, as there was no continuous riverbank. Walcott and his team had to make trails as they went along the canyon walls and cliff sides and up over the ridges that connected one valley to the next. It was slow, dangerous going, where one slip of a mule or man could mean a fall of several hundred feet. Adding to their woes and contrary to Powell's belief, winter did reach the inner canyon. The crew had to fight against wind and snow, and they melted chunks of ice by their campfires to provide water for the animals. It became too much for the assistant fossil collector; too depressed by living in the depths of the canyon and the constant work, he was sent home with Walcott's sympathy and blessing. After more than two months, the remaining crew and animals headed back up the trail out of the canyon and reached a camp on the rim with frostbitten feet.

Walcott had managed to add another 12,000 feet of geological section. Added to his 1879 effort, it made for a "Grand" total of 25,000 feet of stratigraphy—probably the largest section ever measured by a single geologist. But unlike his first trip, this one yielded very few fossils from the massive amount of rock he surveyed. His superior wrote to a British colleague about the mysterious emptiness of the fossil record below the Cambrian Tonto Group:

> Last summer a horse trail was built into the head of the canon & Mr. Walcott a young paleontologist & splendid fossil hunter went into it . . . he found an abundant fauna most decisively or unquestionably Cambrian . . . But what are the beds below? There are over 12000 feet of them . . . Anywhere Walcott cannot find fossils I pity anybody else who tries.

The absence of fossils below the Cambrian was a mystery not just of the canyon but of the whole planet. It was a mystery that confounded and troubled Darwin but one to which Walcott, though he did not realize it at first, held a few precious clues.

Darwin's dilemma

Darwin was well aware of the absence of fossils below the Cambrian, and he candidly addressed the enigma in *The Origin of Species* when he discussed the "Imperfection of the Geological Record":

> There is another . . . difficulty, which is much more serious. I allude to the manner in which many species in several of the main divisions of the animal kingdom suddenly appear in the lowest known fossiliferous rocks . . . If the theory be true, it is indisputable that, before the lowest Silurian or Cambrian Stratum was deposited a long period elapsed, as long as, or probably far longer than, the whole interval from the Cambrian age to the present day; and that during these vast periods the world swarmed with living creatures. . . . To the question why we do not find rich fossiliferous deposits belonging to these assumed earliest periods, I can give no satisfactory answer.

But, Darwin added, "We should not forget that only a small portion of the world is known with accuracy."

Figure 6.3 The Grand Canyon. A view of the Colorado River from Nun-ko-weap. Walcott had to navigate the steep sides of valleys and canyons such as this. Photo by Mike Quinn, courtesy of the U.S. National Park Service.

The vexing problem for Darwin and all proponents of evolution was that trilobites and other creatures appeared suddenly, as an evolutionary explosion in the Cambrian fossil record. It appeared as if the Cambrian explosion marked the dawn of life and that complex animals emerged in a short span of time without any simpler forms preceding them. This pattern in the fossil record was, of course, disconcerting, as it did not fit the pattern of gradual descent from simpler ancestors, and if left unexplained, Darwin admitted, "may be truly urged as a valid argument" against his theory.

Darwin knew, just as Walcott confirmed in the Grand Canyon, that below the lowest fossil-bearing layers of the Cambrian, in which shelly creatures and trilobites were abundant, there was a vast expanse of rock that was apparently empty of life. In his official scientific report, Walcott described the 12,000 feet of rock where "ripple marks and mud cracks abound in many horizons, but not a trace of a fucoid [brown algae], or a molluscan or annelid trail was observed."

He reported that he found only a few bits of other fossils, about which he was not enthused: "But for the discovery of a small Discinoid shell, a couple of specimens of a Pteropod allied to *Hyolithes triangularis*, and an obscure Stromatopora-like group of forms, the two and a half months search for fossils would have been without results." Accustomed to finding hordes of specimens, he was understandably underwhelmed.

Walcott first attributed these few fossils to the Cambrian and asserted that they "could scarcely have been the only representations of the life in the sea at the time."

But, happily, Walcott was off the mark. In other places he had been in North America, the Cambrian was underlain by so-called Archean rocks, which were obviously greatly altered. However, in the Grand Canyon he had found a vast section of typical-looking sedimentary rock below the lowest Cambrian, trilobite-bearing layer. He initially assumed that the thousands of feet of rock were, based on their appearance, also Cambrian. But he was wrong. Several years later, he realized that that vast section of sedimentary rock below was laid down during an immense period of time before the Cambrian: It had to be of Precambrian age. The few fossils he had found in it were the first clear evidence of Precambrian life. Life did not begin all of a sudden in the Cambrian.

The "stromatopora" that he first found so unimpressive are today called "stromatolites," layered conical or spherical structures formed by mats of single-celled blue-green bacteria (cyanobacteria). Those small "discinoid" fossils also proved to be bona fide Precambrian life. They were found within the so-called Chuar Group of rock strata, and Walcott interpreted them as the compressed remains of a shelly creature such as a brachiopod and dubbed them *Chuaria circularis*. They turned out to be the remains of very large spherical algae. While they weren't brachiopods, Walcott is nonetheless credited with finding the first cellularly preserved Precambrian organisms.

The great void of the Precambrian record was finally being filled. We now know that that record extends back some 3 billion years, which is, as Darwin suspected, a period far longer than the interval from the Cambrian to the present (543 million years).

Snowshoe Charlie

It was a very long time before Walcott returned to the Grand Canyon. There was a lot more geology to know, and during the next decade he roamed all over North America—to Vermont, Texas, Utah, New York, Massachusetts, the Canada-Vermont border, North Carolina, Tennessee, Quebec, Colorado, Virginia, Alabama, Pennsylvania, Maryland, New Jersey, Montana, and Idaho—with almost all his fieldwork focused on the Cambrian.

In the midst of it all, he also managed to woo and marry Helena Stevens in 1888. They had quite a honeymoon, or as Walcott's biographer Ellis Yochelson put it: "otherwise known as fieldwork." After a short visit to Montreal, the couple headed to Vermont, where Charles showed Helena some of the romance of field geology; they then went on to Newfoundland and collected trilobites together. In one three-week span, they filled ten barrels and two boxes with fossils. After six weeks they sailed to England and spent several weeks in Wales, the "birthplace" of the Cambrian. The long geo-honeymoon was a great success both for geology and their family, as their first son was born the following May.

A man with so much field experience was invaluable to the USGS, and its leaders began to solicit Walcott's advice in guiding its missions. Walcott was put in charge of all paleontological work in 1892 and promoted to geologist in charge of Geology and Paleontology in 1893. During this time, Director Powell and Congress had a falling-out over water policies in the West and the direction and management of the USGS. Walcott became involved in explaining the agency's work and priorities to legislators on Capitol Hill. He always shot straight and showed such a command of so many fields and issues that the politicians grew to trust and respect him.

In the spring of 1894, Powell resigned, and President Grover Cleveland nominated Walcott as the new director of the USGS, and the Senate promptly confirmed him. Walcott whipped the USGS into shape and regained Congress' support. In his mid-forties, with three young children at home, the top position seemed to be the capstone on a hard-earned career and the end to all his wanderings.

Walcott, however, was just getting started. With the USGS so intimately involved in the exploration and management of the country's resources, he became a major figure on the scientific and governmental scene in Washington. He was promptly elected to the National Academy of Sciences and made an acting assistant secretary of the National Museum (which became the Smithsonian). He was adept at navigating federal agencies, Congress, and even the White House. When a furor erupted over the setting aside of forest reserves by the lame-duck President Cleveland, Walcott stepped in. He approached a pivotal senator (by chance, a neighbor), drafted a new bill, and lobbied the secretary of the interior and President McKinley. The senator introduced the bill, and the forest reserves were saved.

Walcott plied his political talents to the benefit of many organizations. The National Museum badly needed a new building, but any appropriation had to have the support of the Speaker of the House. Knowing that the Speaker often took walks down Pennsylvania Avenue, Walcott would glide by in horse and buggy and offer him a jaunt through a nearby park. Walcott never talked business, but the Speaker got the message, deciding: "Walcott, you may have a building for the survey or one for the National Museum, but you can't have both." Walcott went for the museum.

When Andrew Carnegie was stifled in his bid for a congressional charter for a Carnegie Institution of Washington, Walcott smoothed the way and became a co-founder of this eminent research institution. His quiet way of getting things done earned him the nickname "Snowshoe Charlie," for his adept maneuvering left no tracks.

It was just the sort of skill that presidents needed for sensitive topics. When Theodore Roosevelt took on irrigation and water projects (such as dams and levees) as a major issue of his administration, he wanted it under the control of Walcott, who had been "tested and tried and we know how well they [the USGS] will do their work." Roosevelt's confidence in Walcott only grew as the director fended off the special interests who wanted water projects for their states. Roosevelt soon asked Walcott to take on more tasks, such as the reorganization of all of the branches of government dealing with science. But their passions converged most strongly on the matter of conservation, and Walcott was frequently summoned to the White House for conferences about forests, rivers, parks, and politics.

The exploration and subsequent exploitation of the vast timber, mineral, and energy resources of the West raised a very thorny issue: How could they protect the western lands and wildlife from decimation? Concern that the United States lacked adequate laws to protect prehistoric artifacts, cliff dwellings, cemeteries, caves, mounds, etc., as well as areas of scientific or scenic value, the Antiquities Act of 1906 was forged. It empowered the president to set aside as a national monument or park whatever lands he deemed necessary for their preservation. The act was a precursor to the establishment of the National Park Service. One of the first areas Roosevelt designated as a national monument was the Grand Canyon (1908).

By 1907, Walcott was turning fifty-seven and had spent thirteen years at the helm of the USGS. It was still not a time to be thinking about retirement, and Walcott showed no signs of slowing down whatsoever. During his entire tenure and despite all of his other responsibilities, Walcott managed to escape Washington for the field on a regular basis, particularly in the summers. He made several expeditions each to Montana, Utah, and Nevada. Once the children were just a bit older, these adventures came to be family affairs with the kids joining the treasure hunts. Nothing reinvigorated Walcott like the fresh air of the West, a good campfire, a ride on a sturdy horse, and—oh yes, a barrel of trilobites. When in Washington, he continued to work on his fossils in a side room off of his office.

At a time when most careers would be winding down, particularly one that had been so physically demanding, Walcott was still open to new challenges. In late 1906, the secretary of the Smithsonian Institution passed away and the top post was vacant. The Regents wanted

Walcott to take over. Roosevelt was reluctant to let him leave the USGS, but he ultimately relented, and the Regents got "a man who was not only a recognized scientist, but a man of decided executive ability."

Roosevelt gave a White House dinner in Walcott's honor.

The big bang

Walcott's new job did not change his work habits. The summer field trips continued, and the Smithsonian's staff soon learned that the secretary was not to be disturbed between ten in the morning and two in the afternoon, when he worked on his trilobites and other fossils.

Nor did his new appointment change his cozy relationship with Roosevelt. Walcott continued to be invited to the White House for consultations, especially on matters of conservation. Roosevelt, in turn, was a solid supporter of the Smithsonian. In 1908, he informed Walcott of a great plan he was hatching. As soon as his term was over, in March 1909, he and his son Kermit were going on a long safari and hunting expedition across Africa. Roosevelt wrote: "Now it seems to me that this opens the best chance for the National Museum to get a fine collection of not only the big game beasts, but of the smaller animals and birds of Africa, and looking at it dispassionately I believe the chance ought not to be neglected." Walcott agreed. The trouble was, while Roosevelt's and his son's expenses were covered, the president couldn't afford the expenses of a taxidermist and of shipping the specimens. Nor could the Smithsonian. "Snowshoe Charlie" went into action. Unbeknownst to the Smithsonian staff, he quickly raised $40,000, a small fortune at the time, from private donors.

Roosevelt, determined to make his a scientific expedition, prepared himself by reading five books a week on African wildlife, to the point that he was as knowledgeable as any naturalist on the planet. The preparation paid off. From his safari from Mombasa to Khartoum, the National Museum received nearly 12,000 mammals, birds, reptiles, and amphibians, the quality of which surpassed that of many European museums. Walcott was on one of the ships that steamed out of New York harbor to greet the former president on his return.

But Roosevelt was not the only one to return triumphant from the field that year. Walcott spent the second half of the summer in the Canadian Rockies, as he had the two previous summers. He, Helena, and their thirteen-year-old son, Stuart, ventured to Alberta by train and eventually moved on to Field, British Columbia. From there, they rode their horses high up into the Yoho Valley. The dramatic scenery was often eclipsed by the equally dramatic weather, with thunderstorms, snow squalls, and sleet forcing them to take shelter.

In the last days of August 1909, they made their way up Burgess Pass and dismounted to go collecting. As they inspected some loose blocks of shale that had been brought down by a snow slide, they saw beautifully preserved crustaceans—a lot of them. Walcott had never seen such fine specimens, or the species (Figure 6.4), and they hauled many samples back to camp. Luckily, the good weather held another day, and they returned to the site to find equally fine, well-preserved sponges inside the slabs. The fossils were so remarkable, Walcott was eager to find the source of the fallen rock, so he scampered up the slope. He did find more fossil-bearing slabs, but he could not find the original beds from which the slabs had fallen before the weather turned nasty. Walcott knew he was on to something extraordinary but had to retreat for the season.

The next summer, in July 1910, Walcott, Helena, and two of their sons, Stuart and Sidney, along with a pack handler, Jack Giddie, eagerly headed back up the Yoho Valley. Jack and Stuart rode ahead and reported back that the campsite, near Burgess Pass, was still under four feet of snow, delaying them for three weeks. Once the snow melted, they wasted no time working their way up the 1,200-foot slope above camp to start collecting. They quickly found a large number of what Walcott called "lace crabs," less-than-inch-long crustaceans that he dubbed "Marrella" in honor of a colleague (Figure 6.5). They found an abundance of trilobites, too, but they also found a lot of what Walcott described as "odds and ends," code words for creatures that baffled his expert eye.

They looked through every layer of limestone and shale in search of the fossil-bearing band. At last they found it—a vein about 7–8 feet thick and about 200 feet long—and the operation went into high gear. For thirty days they quarried the shale. Walcott had learned how to use dynamite as a youth on Rust's farm, and it was a handy way to remove a lot of rock from Burgess Pass. He and the boys slid

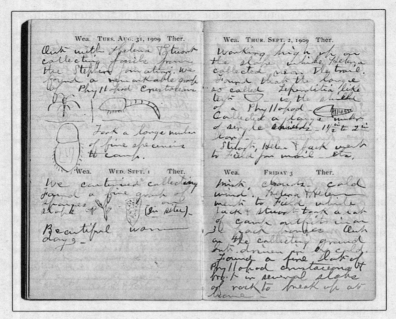

Figure 6.4 Eureka! The pages of Walcott's field diary on the days surrounding the discovery of the Burgess Shale fossils. The drawings on
August 31 depict a couple of crustaceans and a trilobite, and those on
September 1, a sponge. From the Smithsonian Institution Archives, used
by permission.

the blasted blocks down the slope to the trail, where they loaded them
onto packhorses to carry them down to camp. There, Helena led the
operation of splitting the shale, trimming fossil-containing rocks, and
packing them for transport to the railway station at Field, some 3,000
feet below the camp. The operation flowed with the weather. On good
days they blasted and hauled; on stormy days they stayed in camp
and continued to split the shale and discover the critters inside.

After a month, Walcott was exhausted but elated. He wrote to a
colleague at the Smithsonian: "The collection is great . . . It has more
fine, new things than I should have supposed could be found."

What he'd found was vivid evidence of a much more diverse animal kingdom in the Cambrian than had ever been seen or imagined.
The Burgess Shale contained much more than the trilobites and brachiopods of other Cambrian deposits. Its exquisite quality also pre-

served soft-bodied animals (without shells or hard outer skeletons), which represented many of the other major divisions of the animal kingdom (phyla), including annelid worms, priapulids, lobopodians, and even a chordate (Figure 6.5). The arthropods, a well-established Cambrian group thanks to the trilobites, were the most numerous animals in the shale, but they, too, were remarkable for the diversity of forms represented. Walcott gave colorful names to many of the creatures. One arthropod, for example, was called *Sidneyia inexpectans* (Sidney's discovery) in honor of his fourteen-year-old son's discovery of the type specimen (Figure 6.5).

The diversity of animal forms was both stunning and baffling. Bizarre creatures, such as the five-eyed *Opabinia* and the largest fossil, *Anomalocaris*, were also named by Walcott, who took his best shot at classifying them, assigning them to new arthropod orders or families within existing classes. Subsequent research has revealed them to be related to, but different from, any living classes of arthropods.

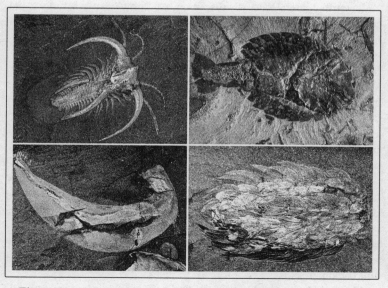

Figure 6.5 An assortment of Burgess animals. Top: *Marrella* is a crustacean; *Sidneyia* is a large arthropod. Bottom: *Ottoia* is a soft-bodied member of another phylum, the priapulids; and *Wiwaxia's* classification is still debated. Photos courtesy of the Smithsonian Institution.

Other Burgess animals, such as *Wiwaxia*, have defied firm classification to this day (Figure 6.5).

Regardless of their exact taxonomy, the significance of these fossils is manifold. The Burgess Shale offered the first glimpse of a riot of life forms in the Cambrian seas. Nothing like it had been found before, and few locales since have been found to rival it. For many of the soft-bodied animals, the Burgess Shale was not just their earliest appearance in the fossil record but their only fossil record at the time. Their discovery pushed the origins of most modern animal phyla back in time to at least the Cambrian. With so many phyla making their first appearance in the Cambrian, later scientists came to call this period the "Big Bang" of animal evolution.

These spectacular fossils have raised many questions for paleontologists, geologists, and biologists for a century. Foremost among them are: What triggered the Cambrian Explosion? Why was life generally small and simple for so long, and why and how did large, complex forms seemingly burst on the scene?

A century after Walcott's discovery, we know quite a bit more, but certainly not all we would like to about the Cambrian Explosion. Our most solid set of facts concern the time frame of the "Explosion," which was completely beyond the reach of methods available in Walcott's day. He thought that the fossils were 15–20 million years old. We now know that the Cambrian period began approximately 543 million years ago and that the Burgess fossils are about 505 million years old.

Other, older Cambrian deposits have been discovered that also contain a great variety of animal forms—in Greenland (Sirius Passet; c. 518 million years old) and, most recently, in China (Chengjiang; c. 520 million years old). While the fossil record of sponges and cnidarians (e.g., jellyfish) extends into the Precambrian, clear body fossils of other animals are generally absent before the Cambrian. The fossil record suggests that these other animal groups began emerging about 530 million years ago, and thus the explosion was under way for some 20–25 million years before the Burgess animals were buried and became fossils.

Altogether, the unfolding of the Cambrian took perhaps 40–50 million years. Such a long time might stretch the connotation of "explosion." But compared to what existed earlier, the Cambrian marks a dramatic change in life forms; as my colleague and the Harvard paleontologist Andy Knoll has underscored, "Those 50 million years reshaped more than *3 billion* years of biological history." The explosion was real, but not a sudden event.

What ignited the explosion? There was a dramatic rise in oxygen levels in the late Precambrian seas. Since larger creatures need to distribute oxygen to cells deep inside their bodies, some scientists favor the idea that rising oxygen levels in the ocean made possible the emergence of larger animals. These animals may then have triggered an explosive ecological competition between predators and prey that rapidly filled the seas with all sorts of creatures.

Legacy

Walcott returned to the Burgess quarry for several more seasons after 1910 and to the Canadian Rockies every year through 1925, when he was seventy-five. His beloved Rockies and the heavy work were all too often needed therapy for a very heavy heart. Just before the 1911 field season, Helena was killed in a train accident. In 1913, his eldest son, Charles, died, and in 1917, Stuart died; he was the first American aviator killed in World War I. Walcott wrote to President Woodrow Wilson that in such times of grief, "steady, systematic work is one's salvation." He lived by these words, for he managed to bring back an astonishing horde of more than 65,000 Burgess specimens to the Smithsonian Institution; they are now a national treasure.

Walcott did not neglect the other collections or missions of the national museum during his twenty-year administration. He helped to persuade the Detroit industrialist Charles Freer to establish what has become the renowned Freer Gallery of East Asian Art at the Smithsonian. As secretary of the institution whose charter was to promote "the increase and diffusion of knowledge," Walcott also had a keen interest in the development of American aviation. It was in his role as secretary that he presented a medal to the Wright brothers that February day in 1910 (Figure 6.1). He was later pivotal in getting Congress, the military, and President Wilson to establish a National Advisory Committee for Aeronautics (NACA) in 1915, whose executive committee he chaired. Forty-three years later, in 1958, as the country rushed to catch up to the Soviet Union after the launch of

Figure 6.6 The Burgess Shale Quarry. Walcott at the quarry he established. Photo courtesy of the Smithsonian Institution.

Sputnik, this organization grew into the National Aeronautics and Space Administration (NASA) and was given the charge to explore outer space.

From the depths of the Grand Canyon to the top of the Canadian Rockies, from the microscopic relics of the Precambrian to the animals of the Big Bang, from the opening of the American West to the opening of the space race, from Trenton Falls to the White House, Snowshoe Charlie left some very deep tracks.

Figure 7.1 Roy Chapman Andrews, explorer. Approaching a kite's nest in the badlands of the Gobi desert of Mongolia. The ranger hat and pistol on his side were his trademarks when in the field (image #410988; photo by James Shackelford; American Museum of Natural History Library).

❧ 7 ❧
Where the Dragon Laid Her Eggs

> Dreams come true. Without that possibility,
> nature would not incite us to have them.
> — *John Updike*

GROWING UP IN THE LATE 1800s in Beloit, Wisconsin, a small industrial town on the banks of the Rock River, young Roy Chapman Andrews spent every minute he could out of doors, whatever the weather. When confined indoors, he made his mother read *Robinson Crusoe* to him again and again, even though he knew it by heart. He dreamed of living on a deserted island and fending for himself.

By the time he was eight, Roy was passionate about nature. He wandered through the woods with binoculars, a notebook, and field guides to birds. Inspired by visits to the Field Museum of Natural History in Chicago, he put together a little museum in the attic of his house, where plants, minerals, fossils, stuffed animals, and other artifacts were carefully labeled and displayed.

His parents indulged their son's enthusiasm. On his ninth birthday, his father gave him a small single-barrel shotgun. Roy took his new prize goose hunting and stalked several birds that were floating on the edge of a marsh. Crawling on his stomach through mud and water, the boy closed in and fired. Three of the geese slowly collapsed with a gentle hissing sound—he had shot someone's decoys full of holes!

Their owner, Fred Fenton, jumped out of the bushes and was not at all pleased. Roy was scared to death and took off for home, sobbing. Roy's dad, however, roared with laughter. It turned out that he hated Fenton, and he promised to buy Roy a double-barrel shotgun "so you can get 'em all next time." Roy got the new weapon within a week.

Roy loved to camp, to fish, and to hunt, often on his own. One evening, after his camp dinner of bread and bacon, he curled up to sleep in the open under a great tree. During the night, he felt something wriggling in his hair and brushed at it sleepily with his hand. A cold body wrapped around his wrist and arm. Roy yelled out and shook with fright. It was a harmless garter snake, but the incident haunted him for weeks and a lifelong hatred of snakes was cemented.

Roy's pursuit of game led to an interest in taxidermy. From a book, he taught himself how to mount birds and other animals. Soon he began to mount birds and deer for neighboring hunters. He always had plenty of work during the fall shooting season—and plenty of money each Christmas. His earnings helped him pay his own way through the excellent Beloit College nearby.

Though he worked hard on his taxidermy, he was not a very diligent student. He was bored by most subjects other than science and did very poorly at math. His appetite for exploration and natural history was fueled by accounts of the great African explorers, Richard Burton, John Speke, David Livingstone, and Henry Stanley, and by his many romps around the Wisconsin countryside.

Roy did take advantage of one important opportunity at college—to meet professional scientists. One day he cornered a geologist from the American Museum of Natural History who had come to Beloit to lecture. Roy explained his passion for natural history and persuaded him to drop by a tavern to see some of Roy's mounted heads. Impressed, the geologist advised Roy to write to the museum's director about the possibility of a job.

Roy wanted to be a naturalist and explorer so badly that no other profession ever entered his mind. Though he had never before traveled farther than the 90 miles to Chicago, on the day of his graduation he announced to his parents that he was going to New York to try to get a job at the natural history museum.

A whale of a start

The day after he arrived in New York, Roy made his way to the museum for an appointment with Dr. Bumpus, the director. He nervously answered questions before Dr. Bumpus gently explained that there were no positions open. Roy's heart sank, but then the entrepreneur rose within him, and he blurted out, "I'm not asking for a position. I just want to work here. You have to have someone to clean the floors. Couldn't I do that?"

Dr. Bumpus said, "A man with a college education doesn't want to clean floors!"

"No," Roy said, "not just any floors. But the Museum floors are different."

Impressed by Roy's spark, Dr. Bumpus hired him and took him out to lunch. Little did Dr. Bumpus know that Roy would become the most famous Museum employee, and one day hold his job.

Assigned to the taxidermy department, Roy began each day by mopping the floors, just as he was hired to do, but he was soon given more interesting assignments. His first big opportunity came when Dr. Bumpus made him an assistant to a scientist who was to build a life-size model of a whale. Roy and a friend figured out that the best way to do it was with wire netting and papier-mâché.

Roy had never even seen a whale. Soon thereafter, a whale carcass washed up on Long Island, and Roy and his model-building friend got the assignment to bring back the whole skeleton. It was an enormous task to free the skeleton from the mass of meat and blubber while the rotting carcass sank into the sand and was pounded by the surf. Soaked, frozen, and exhausted, they worked for days until they were able to secure this novel prize for the museum. Proud of his young protégé, Dr. Bumpus introduced Roy to many of the important visitors to the museum, from leading naturalists to such industrial tycoons of the early 1900s as Andrew Carnegie.

Roy's growing reputation concerning whales led to his first real field expedition. Little had been documented about whale behavior, so Roy coaxed the director to let him accompany whaling ships operating off Vancouver. Even though he was tortured by constant seasickness, Roy was able to observe whales up close, including the birth of a calf. His success on that first mission led to many more whale studies in different parts of the world over the next eight years, particularly in the Far East off Japan, Korea, and China. During this time, Roy developed a strong attachment to the Orient and expanded his reputation as a naturalist. For example, he "rediscovered" the gray whale, once thought to have been hunted to extinction. Most important, Roy's many expeditions and frequent side trips provided opportunities for adventures that fueled his appetite for exploration, as well as his self-confidence.

Once, when Roy had a few weeks' delay, waiting for a ship out of the Philippines, he had a steamer take him and two Filipino assistants to an uninhabited island. On this palm-fringed paradise, surrounded by coral reefs, Roy lived the *Robinson Crusoe* experience he had always dreamed of. Perhaps it was a bit too authentic: the ship left enough food for just five days, and a broken propeller delayed its return by over two weeks.

When the food was gone, Roy didn't care—he was too happy wading in the tidepools, catching fish and crabs, and sleeping under the stars. He built a snare to trap birds and lived as a castaway. When the ship finally showed up, Roy's heart was heavy, for he feared he would never again be so content as he was on "his" deserted island.

But he need not have worried, much greater adventures lay ahead.

Asia dreaming

After eight years spent chasing whales around the world, Roy wanted to get onto land. But where were the greatest opportunities for discovery?

Roy was smitten with Asia, China in particular. From the moment he entered Peking (now Beijing), he loved the colorful dress of the people and the stone walls, battlements, and statues that were steeped in history. And he was awed by the Great Wall. Going to see it for the first time, he walked toward it head down, so that when he looked up he would see all of its majesty at once: "At last, there it was spreading its length like a slumbering gray serpent over the hills, into the valleys, and up the sides of precipitous peaks as far as my eyes could reach. No other sight on earth has ever stirred me as did the Great Wall of China. . . . I knew that some day I should return."

The Great Wall had been built to protect Northern China from the Mongols and other invaders. Roy would pass through its gates again and again to explore the vast region beyond.

Looking for a reason to explore Asia, Roy found it in the theories of the new president of the museum, Henry Fairfield Osborn. Osborn was an eminent paleontologist who had led fossil expeditions in the

American West and described and named *Tyrannosaurus rex*. He believed that Asia was the home of ancient humans and the origin of much of the animal life of Europe and America. Testing these ideas required a much greater knowledge of the living and fossil fauna of Central Asia than was available at the time. What better justification for an expedition could there be than to test his boss' favorite ideas? Roy proposed to mount an expedition to gather the necessary scientific data. Osborn was, of course, quite enthusiastic.

In 1916, Roy made his first collecting foray to the Yunnan province of southeastern China. He came back with preserved specimens of 2,100 mammals, 1,000 birds, and many fish and reptiles, all from regions never before seen by a collector.

While he was in Yunnan, the United States entered the First World War. Through a friend in Naval Intelligence, Roy secured an assignment as a spy, working under the cover of collecting zoological specimens. This allowed him to stay in Asia for the duration of the war, traveling throughout China, Manchuria, Mongolia, and as far as Siberia. Roy loved the "work," as he saw "new country, new customs, new people everyday." He also filed intelligence reports on political developments and military matters. Most important, his travels allowed him to scout out promising places for future exploration.

He was captivated by Mongolia and the Gobi desert. He made his first trips by motorcar, a novelty at the time and a much more comfortable, and speedier, mode of transport than a camel. After passing through the Great Wall at Kalgan, he made his way toward Urga, the capital of Mongolia. Roy was thrilled by the colorful canyons, mountains, ravines, and the vast plains beyond; he wrote:

> Never again will I have such a feeling as Mongolia gave me. The broad sweeps of dun colored gravel merging into a vague horizon; the ancient trails once traveled by Genghis Khan's wild raiders; the violent contrast of motor cars beside majestic camels fresh from the marching sands of the Gobi! All this thrilled me to the core. I had found my country. The one I was born to know and love.

Just after the war, in the spring of 1919, Roy made the museum's first expedition into Mongolia and the Gobi. It was strictly a zoological collecting trip, albeit in an almost entirely unknown country.

Roy's efforts produced another 1,500 mammal specimens for the museum and a plan for a great expedition, the likes of which had never been imagined before.

The big plan

Roy went to New York to see Professor Osborn and outline his grand idea. Osborn sensed Roy had a pitch to make, and let Roy spill his plan. Roy said that in order to test Osborn's theories, they "should try to reconstruct the whole past history of the Central Asian plateau —its geology, fossils, past climate, and vegetation. We've got to collect its living mammals, birds, fish, reptile, insects, and plants and map the unexplored parts of the Gobi. It must be a thorough job: the biggest land expedition ever to leave the United States."

Roy had thought out the special logistical challenges, and his experience in motorcars was crucial. He told Osborn he could cover far more ground if, instead of using camels, which could cover only 10 to 15 miles a day, he used motorcars, which could go 100 miles, and deployed a caravan of camels to act like supply ships to the fleet of cars. By sending the camels ahead of the fleet in winter, with food and equipment, the caravan and the cars could then rendezvous deep in the desert in the summer.

Osborn peppered Roy with questions, but Roy had anticipated all the details. Most important, Roy impressed on Osborn the importance of taking along the best experts in many fields—including geology, paleontology, and mammalogy.

Osborn was sold: "Roy, we've got to do it. This plan is scientifically sound. Moreover, it grips the imagination. Finances are the only obstacle."

Roy thought he needed $250,000 (in 1921 dollars; perhaps $10 million today) for a five-year expedition. He had a plan for that, too. He had rubbed shoulders with enough business titans to realize that they might support such a venture if it brought them some prestige in high society.

The Beloit taxidermist was right again.

He first visited the banker J. P. Morgan. When Roy spread out his map, Morgan listened with rapt attention. Roy laid out his whole plan in fifteen minutes. When he stopped, out of breath, Morgan blurted: "It's a great plan; a great plan. I'll gamble with you . . . All right, I'll give you fifty thousand. Now you go out and get the rest of it." Morgan sent Roy to a fellow banker at Chase National Bank, who ponied up $10,000. Other members of the New York elite, including John D. Rockefeller, followed suit. Roy enjoyed his adventures on Wall Street; he liked the gambling spirit of the titans of steel, oil, railroads, banks, and other industries.

The expedition soon captured the attention of the newspapers and the general public. Roy was deluged with thousands of letters from people volunteering to join the expedition, many from teenage boys. But he had already selected his team very carefully.

It included Walter Granger, a paleontologist and second in command; Charles Berkey, the chief geologist; Clifford Pope, a herpetologist; Bayard Colgate, the chief of motor transport; Frederick Morris, a geologist; and J. B. Shackelford, a photographer (Figure 7.2). These

Figure 7.2 The 1922 Expedition Team at Tsagon Nor, Mongolia. Second row, left to right: Morris, Colgate, Granger, Badmajapoff, Andrews, Berkey, Larsen, Shackelford. Top row: Chinese technical and camp associates. Bottom row: Mongolian interpreters and caravan men. From *The New Conquest of Central Asia: A Narrative of the Explorations of the Central Asiatic Expeditions in Mongolia and China, 1921–1930,* by Roy Chapman Andrews (1932). The American Museum of Natural History, New York.

individuals were well chosen. They got along exceedingly well, and in the field their strengths complemented one another.

With the money and the team falling into place, the major task of preparation came next. Roy put his considerable experience into the selection of the equipment and food and the overall planning. He knew that it would be impossible to find anything but meat in the desert, so he ordered great quantities of dried fruits and vegetables—onions, tomatoes, carrots, beets, and spinach from America; the rest of the food was provided by a Marine Corps detachment in China. He obtained Mongol tents and fur sleeping bags, as the nomads knew best how to cope with the desert weather. And he selected the vehicles as well— three Dodge Brothers cars and two one-ton Fulton trucks. The Standard Oil Company of New York donated 3,000 gallons of gasoline and 50 gallons of oil. Altogether, eighteen tons of equipment and supplies were shipped to Peking; they needed seventy-five camels to carry it all. The expedition's headquarters was established in Peking; camels were purchased, loaded, sent out in advance; and finally, on April 21, 1922, the team set out in five cars from Kalgan, passed through the Great Wall, and entered the vast territory beyond (Figure 7.3).

There were many unknowns. Foremost among them was whether they could find any fossils. Professor Osborn had noted, and Roy also knew, that the only fossil known from the central Asian plateau was a single tooth, which had been discovered by a Russian in the late 1890s. The expedition had been ridiculed by some as pointless; they said that the desert was a wasteland of sand and gravel and that Roy might as well search for fossils in the Pacific Ocean. Roy was also told that it was criminal to waste the time of such eminent geologists as Berkey and Morris in a country where the "geology was all obscured by sand."

But Roy's team soon proved the skeptics wrong. Oh, boy, were they ever wrong.

Iren Dabasu

The car caravan quickly made its way toward the Gobi. Making and breaking camp was a fluid operation. The tents were pitched and a fire was started within thirty minutes of selecting a spot. Each team member then took care of his particular tasks. The chief of motor

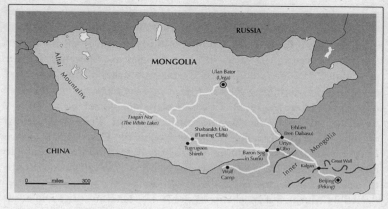

Figure 7.3 Map of the Central Asiatic Expeditions. White lines trace the various routes taken in expeditions between 1921 and 1930. Some of the place names have changed over the years, so both current and former names are shown. Based on a map in L. Rexer and R. Klein (1995), *125 Years of Exploration and Discovery*, Harry N. Abrams in association with the American Museum of Natural History, New York. Redrawn by Leanne Olds.

transport filled the gas tanks and thoroughly inspected each car for loose bolts or cut tires. The geologists transcribed their notes from the day's work. The photographer loaded his film magazines and made a diary of the day's photographs. The taxidermists set traps for mammals. And, if there was a rock outcrop or exposure nearby, the paleontologists would make a quick search for fossils.

Four days into the expedition, Roy and his staff set up camp at Iren Dabasu while Berkey, Morris, and Granger stopped a few miles away to scout for fossils. As Roy was enjoying a beautiful sunset over the desert, Granger and the geologists roared into camp. Granger's eyes were shining as he reached into his pockets and held out several fossils, exclaiming, "Well, Roy, we've done it. The stuff is here. We picked up fifty pounds of bone in an hour."

They would identify some of the teeth as from a rhinoceros, but they weren't sure about the other mammal fragments. No matter. Everyone was happy and eager to go fossil hunting at first light.

The next morning Berkey came to breakfast with his hands full of fossils; they had camped right on top of another exposure. Granger was puzzled by one leg bone—it was not mammalian. He went to the outcrop where Berkey had found it and exposed a great bone perfectly preserved in the rock—it had belonged to a dinosaur. Berkey proclaimed, "We are standing on Cretaceous strata of the upper part of the Age of Reptiles—*the first Cretaceous strata and the first dinosaur ever discovered in Asia north of the Himalaya mountains.*"

All of Roy's optimism was instantly validated. They were sitting on top of rich mammal and dinosaur fossil beds. There was far more to investigate at Iren Dabasu, but they had no time to celebrate or prospect further. They had to keep moving to stay on schedule with the camel caravan, 350 miles ahead of them.

Led by Merin, a remarkable Mongol, the caravan was to meet up with the cars on April 28. Sure enough, as Roy approached the rendezvous point, he saw the American flag flying from one of the loads. After a thirty-eight-day trek, Merin had arrived at the designated spot one hour earlier. The sight of the camel caravan, spread out in single file among the rocks, was majestic (Figure 7.4).

As the expedition made its way across the desert, it repeatedly faced a stubborn enemy—sandstorms. Roy described the power of and confusion sown by one of the many great storms:

> Slowly I became conscious that the air was vibrating to a continuous roar, louder every second. Then I understood. One of the terrible desert storms was on the way. The shallow basin seemed to be smoking like the crater of a volcano—yellow "wind devils" eddied up and swirled across the plain. To the north an ominous tawny bank advanced at horse race speed. I started back toward camp, but almost instantly a thousand shrieking storm demons were pelting my face with sand and gravel. Breathing was difficult; seeing impossible.

The wind blew for ten days, until the team's nerves were worn thin. But once the storm passed, they again enjoyed the scenery of fantastic sand dunes, red sunsets against lavender mountains, and the fossil treasures poking out of the ground.

Shackelford had an uncanny knack for spotting fossils. One day, near the shore of a desert lake, he actually stumbled over a huge leg bone that had weathered out of a stream bank. The team had been told by some Mongols of "bones as large as a man's body." Here was

Figure 7.4 Expedition caravan. The camel caravan arriving at the Flaming Cliffs in 1925. From *The New Conquest of Central Asia: A Narrative of the Explorations of the Central Asiatic Expeditions in Mongolia and China, 1921–1930*, by Roy Chapman Andrews (1932). The American Museum of Natural History, New York.

proof. It was the upper foreleg (humerus) of a *Baluchitherium* ("the beast of Baluchistar," after the place in Pakistan where in 1913 it was discovered). Roy and Granger later found a skull of the colossal animal—the largest land mammal that ever lived: nearly 17 feet tall, 26 feet long, and weighing 15 tons.

Granger had his hands full with other fossils. In one day alone he found 175 jaws and skulls of various carnivores, rodents, and insectivores. And nearby, he discovered a complete skeleton of a small beaked dinosaur.

As difficult as the desert conditions could be, it was actually fortunate that these fossils were in such remote locations in Mongolia. In China, fossil bones have long been interpreted as those of dragons, which are sacred symbols of power. For at least two millennia, these "dragon bones"—the bones of extinct mammals and dinosaurs—have been collected, ground up into powders, and used in traditional folk medicines. Had the Gobi sites been closer to civilization, they might have been destroyed.

. . .

By September 1, it was time to leave the Gobi. The weather was turning, and great flocks of birds were flying south from the northern tundra. The expedition was low on water, and Roy feared being trapped in a blizzard.

While stopped to look for a well, Shackelford went strolling and found himself on the edge of a vast red sandstone basin. Scampering down the slope, he walked toward a small rock pinnacle supporting a white fossil bone, just waiting to be plucked. Unfamiliar, it turned out to be the skull of a horned dinosaur, later named *Protoceratops andrewsi* in Roy's honor (Figure 7.5). Shackelford reported seeing other bones, so the team decided to camp.

The next day they discovered that these beautiful badlands "were paved with white fossil bones and all represented animals unknown" to any of the team, but they had to leave further exploration to another season. Inspired by the fiery glow of the rocks in the late afternoon sun, Roy dubbed the spot "the Flaming Cliffs," and the team headed back to Peking.

The scientific results of the expedition had exceeded their great-

Figure 7.5 *Protoceratops andrewsi* exposed at the Flaming Cliffs. From *The New Conquest of Central Asia: A Narrative of the Explorations of the Central Asiatic Expeditions in Mongolia and China, 1921–1930,* by Roy Chapman Andrews (1932). The American Museum of Natural History, New York.

est hopes. They had found complete skeletons of small dinosaurs and parts of larger ones, skulls of mastodons, rodents, carnivores, deer, giant ostrich, and rhinoceros, as well as Cretaceous mosquitoes, butterflies, fish, and more. Shackleford had shot 20,000 feet of film of both the expedition and life in the Gobi. Almost all of their specimens were new to science. Professor Osborn sent congratulations: "You have written a new chapter in the history of life upon the earth."

But the team knew they had only scratched the surface. They started preparing for the next season and their return to the Flaming Cliffs.

Where the dragon laid her eggs

Exactly one year after the first expedition, the team again left Peking for the Gobi. Once camped at Iren Dabasu, Roy and another driver went back to Kalgan for more supplies. They almost didn't make it.

Approaching a deep valley, where he knew Russian cars had been robbed a week earlier, Roy was wary. When he spotted a man on horseback with a rifle, he took out his revolver and pinged a shot nearby to scare him off. But then he saw four armed horsemen ahead. The trail was too narrow for him to turn around, so he gunned the engine and raced down the trail to scare the horses. It worked. The bandits had to hold on tightly as their horses bolted and Roy fired off a few shots. It was, in his words, "great fun."

The team trekked the several hundred miles to the Flaming Cliffs, following the tracks their cars had made ten months earlier. Making camp in midafternoon, they scattered to look for fossils. By nightfall, everyone had his own dinosaur skull.

On the second day a new team member, George Olsen, reported at lunch that he thought he had found some fossil eggs (Figure 7.6). The others gave him a good ribbing, but they were curious enough to follow him back to the spot to see what he was talking about:

Then our indifference suddenly evaporated. It was certain they really *were* eggs. Three of them were exposed and evidently had broken out of the sandstone ledge beside which they lay . . .

We could hardly believe our eyes, but, even though we tried to account for them in every possible way as geological phenomena, there was no shadow of doubt that they really were eggs. That they must be those of a dinosaur we felt certain. True enough, it never was known before that dinosaurs did lay eggs . . . although hundreds of skulls and skeletons of dinosaurs had been discovered in various parts of the world, never had an egg been brought to light.

Dinosaur eggs! Roy later admitted, "Nothing in the world was further from our minds."

The eggs were not the only discovery: "While the rest of us were on our hands and knees about the spot, Olsen scraped away the loose rock on the summit of the ledge. To our amazement he uncovered the skeleton of a small dinosaur lying four inches above the eggs."

It was a type of dinosaur completely new to science. Professor Osborn later conjectured that it was an egg thief caught in the act and called the new species *Oviraptor* (egg seizer).

A few days later, five more eggs were found in a cluster, then another group of nine. In two eggs that had broken in half, they could

Figure 7.6 **The first nest of dinosaur eggs.** Discovered by George Olsen at the Flaming Cliffs in 1923. From *The New Conquest of Central Asia: A Narrative of the Explorations of the Central Asiatic Expeditions in Mongolia and China, 1921–1930,* by Roy Chapman Andrews (1932). The American Museum of Natural History, New York.

plainly see the bones of dinosaur embryos. Twenty-five eggs in all were found that year and many more in subsequent years (Figure 7.7).

But the eggs were not the end of the treasure hunt. The team uncovered seventy-five skulls within an area of three miles. Finding so many specimens created a problem. The specimens were prepared and packed by wrapping them in cloth soaked in flour paste. In three weeks, however, they had nearly exhausted all of their flour. Roy polled the team: Should they stop working or use the rest of the flour? They were unanimous: "Let's keep the flour for work," leaving them only tea and meat for food.

They also ran out of the burlap used to wrap fossils, so they had to improvise. First, they cut off all their tent flaps, then used their towels and washcloths. Finally they used their clothes—socks, trousers, shirts, underclothes, and even Roy's pajamas. Among the many dinosaurs thus preserved Osborn would identify several new species, including *Velociraptor*, a swift predator with large talons, and *Tarborsaurus*, a close relative of *Tyrannosaurus rex*.

The horde from the Flaming Cliffs filled sixty supply boxes and gasoline tins and weighed five tons. Buried in the massive haul was one small gem, as yet unrecognized: a tiny skull, barely more than an inch long, found in the same Cretaceous layer as the dinosaur eggs. Granger had labeled it "an unidentifiable reptile." However, once unpacked and studied at the museum, it was clear that it was not a reptile but a mammal. It was the most complete specimen of mammal life in the Cretaceous unearthed to date, clear proof that mammals had lived beside dinosaurs. But one specimen gave them just a glimpse of early mammal life. Finding more mammals would be a top priority for the next expedition.

On the trail of ancient mammals

On their third visit to the Flaming Cliffs, in 1925, Roy brought along a letter that he had been carrying for Granger from W. D. Matthew, the museum's curator of paleontology. In it, Matthew explained the importance of that tiny mammal skull and wrote, "Do your utmost to get some other skulls."

After Roy and Granger discussed the matter for a bit, Granger declared, "Well, I guess that's an order. I'd better get busy." He walked out to the Flaming Cliffs and came back with another mammal skull

Figure 7.7 Another nest of dinosaur eggs. Andrews is on the right, rifle handy; Olsen is on the left. The oblong eggs are between them. From *The New Conquest of Central Asia: A Narrative of the Explorations of the Central Asiatic Expeditions in Mongolia and China, 1921–1930*, by Roy Chapman Andrews (1932). The American Museum of Natural History, New York.

within an hour. Later, he and his assistants spent many days in the scorching sun inspecting thousands of little nodules of sandstone for more skulls. It was tedious, grueling work, but they found seven more skulls, most with lower jaws. Granger kept these delicate specimens in his suitcase for safekeeping.

When the expedition was over, Roy carried these treasures to New York and presented them to Dr. Matthew. Subsequent analysis revealed that they included two families of insectivores (Figure 7.8), some of the first "missing links" in the story of mammal evolution. The fossils revealed that before the end of the so-called Age of Dinosaurs, mammals had already split into the two main lines of marsupial and placental forms. The Flaming Cliffs also yielded a fossil representative of the "multituberculates," an ancient branch of mammals and the only major branch to have become completely extinct.

Roy believed that these fossils were the most valuable finds of the entire expedition and that Granger's intense search was "possibly the most valuable seven days of work in the whole history of paleontology."

Snakes on a plain!

Returning east from the Flaming Cliffs, the expedition found an abundance of geologically younger mammals at the other end of the size spectrum. The team unearthed two fossil titanotheres, including one enormous skull of the rhinoceros-like beasts, previously known only from America. At a new camp, no fewer than twenty-seven mammal jaws were exposed in one layer, and more lay underneath. The fossils included one of a strange, claw-hoofed animal and scores of a small hoofed animal known as *Lophiodon* (a relative of tapirs). Roy concluded that the region must have once swarmed with these mammals, just as dinosaurs had also been abundant here.

The lands they explored turned out to be swarming with more than fossils—it was infested with pit vipers. During the day, three snakes were discovered near the tents, and all the members of the team saw snakes while they were prospecting for fossils.

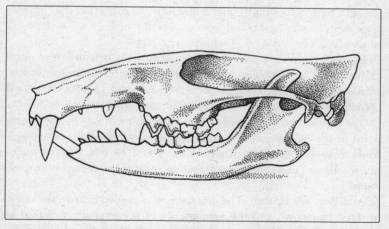

Figure 7.8 An early mammal skull. The skull of *Zalamdalestes*, a shrew-like mammal that lived in the Cretaceous period alongside dinosaurs, was about two inches long. Drawn by Leanne Olds.

That was already far too many snakes for Roy. Then, one night, the temperature dropped to near freezing, and the vipers moved into camp for warmth. All hell broke loose. One of the motor engineers woke in the night and saw a snake near the door of his tent. Checking around his tent, he found snakes around each leg of his cot and one under a gasoline box. Morris, the geologist, cried out, "Dear God, my tent is full of snakes."

The Mongols would not kill the snakes because the camp was in a sacred spot near a temple. With no such restraints, the Americans dispatched forty-seven vipers. Everyone was on edge. Roy leaped when he saw what turned out to be a piece of rope. Granger attacked what was in fact a pipe cleaner. No one was bitten, but after two days of being surrounded by snakes, the team packed up, left Viper Camp, and headed for Peking.

Under a lucky star

When the 1925 expedition left the Flaming Cliffs, Roy noted that these few square miles had given them more than they had hoped for from the entire Gobi desert—the first dinosaur eggs, a hundred skulls and skeletons of new dinosaurs, and eight Cretaceous mammal skulls. Looking at the beautiful red rock formations for what he thought (correctly) might be the last time, he regretted that his caravan would never again "fight its way across the long miles of desert to this treasure house of Mongolian pre-history."

Roy and his team made two more expeditions into different parts of the Gobi, in 1928 and 1930. Wars and sentiments against foreigners precluded expeditions in 1926, 1927, and 1929, and political upheaval eventually ended further fieldwork altogether.

They never did find ancient human remains, the original selling point of the expedition. But their discoveries had produced enough leads to keep many scientists busy for years. To this day, the mammal fossils and dinosaurs from the Gobi are still the subject of intense study.

And Roy was famous. The dinosaur eggs landed him on the cover of *Time* magazine, and the prominence of the expedition, based in New York and backed by such prominent industrialists, ensured widespread press coverage of his exploits. Roy's lectures drew throngs, and he wrote many popular magazine articles and several books about the expeditions. He was given many honors, including medals that had been earned by only the most intrepid explorers — Peary, Scott, Shackleton, Amundsen, and Byrd. And in 1935 Roy became the director of the museum where he once so enthusiastically swept the floors.

In New York, he and his wife traveled in the highest circles, keeping company with fellow explorers such as William Beebe, the aviators Charles Lindbergh and Amelia Earhart, as well as various movie stars. Roy, too, was a movie star of sorts. Featured in newsreels and the papers, usually photographed in his ranger hat with a revolver on his hip, Roy was the image of a new breed of explorer-scientist. If all that along with a hatred of snakes sounds a bit like the character Indiana Jones in recent movies, perhaps it is not a coincidence. George Lucas, who created Indiana Jones, is reported to have been inspired by characters in B-movie serials of the 1940s and 1950s; they in turn were probably influenced by Roy's and others' accounts of his adventures.

In his autobiography, *Under a Lucky Star: A Lifetime of Adventure*, Roy recalled how as a boy he "always intended to be an explorer, to work in a natural history museum, and to live out of doors." He closed by admitting how lucky he had been to be able to live out his dreams, how, for him, "always there has been an adventure just around the corner — and the world is still full of corners!"

Figure 8.1 Luis and Walter Alvarez at limestone outcrop near Gubbio, Italy. Walter Alvarez (right) is touching the top of the Cretaceous limestone, at the K/T boundary. Photograph courtesy of Ernest Orlando Lawrence, Berkeley National Laboratory.

∾ 8 ∾

The Day the Mesozoic Died

The beginning of knowledge is the discovery
of something we do not understand.
— *Frank Herbert, author of* Dune

BUILT ON THE SLOPES of Monte Ingino, in Umbria, the ancient town of Gubbio, founded by the Etruscans between the second and first centuries B.C., boasts many well-preserved structures that document its glorious history. Its Roman theater, Consuls Palace, and various churches and fountains are spectacular monuments to the Roman, Medieval, and Renaissance periods. It is one of those special destinations that draw tourists to this part of Italy.

It was not the ancient architecture (although it certainly made the fieldwork more enjoyable) but the much older natural history preserved in the rock formations outside the city walls that attracted Walter Alvarez, a young American geologist, to Gubbio in the early 1970s. Just outside Gubbio lay a geologist's dream—one of the most extensive, continuous limestone rock sequences anywhere on the planet (Figure 8.1). The *Scaglia rossa* is the local name for the attractive pink outcrops along the mountainsides and gorges of the area. (*Scaglia* means "scale" or "flake" and refers to how the rock is easily chipped into the square blocks used for buildings, such as the Roman theater. *Rossa* refers to its pink color.) The massive formation is composed of many layers that span about 400 meters in all. Once an ancient seabed, the rocks represent some 50 million years of Earth's history.

Geologists have long used fossils to help identify rocks from around the world, as Walcott did in the Grand Canyon, and Walter followed this strategy in studying the formations around Gubbio. Throughout the limestone he found fossilized shells of tiny creatures called foraminifera ("forams"), a group of single-celled protists that can be seen only with a magnifying lens. But in one centimeter of clay separating two limestone layers, he found no fossils at all. Furthermore, in the older layer, below the clay, the forams were much larger than in the younger layer above the clay (Figure 8.2). Everywhere he looked around Gubbio, he found that thin layer of clay and the same distribution of forams above and below it.

Walter was puzzled. What had caused such a change in the forams? How fast did it happen? How long a period of time did that thin layer without forams represent?

These questions, about seemingly mundane microscopic creatures and half an inch of clay in a 1,300-foot-thick rock bed in Italy, might appear trivial. But their pursuit led Walter to a truly Earth-shattering discovery about one of the most important days in the history of life.

The K-T boundary

From the distribution of fossils and other geological data, it was known that the Gubbio formation spanned parts of both the Cretaceous and Tertiary periods. The names of these and other geological time periods come from early geologists' ideas about the major intervals in Earth's history, and from some of the features that mark particular times. In one scheme, the history of life is divided into three eras—the Paleozoic ("ancient life," the first animals), the Mesozoic ("middle life," the age of dinosaurs), and the Cenozoic ("recent life," the age of mammals) (Figure 8.3). The Cretaceous period, named after characteristic chalky deposits, forms the last third of the Mesozoic era. The Tertiary period began at the end of the Cretaceous 65 million years ago and ended at the beginning of the Quaternary period 1.6–1.8 million years ago.

Walter and his colleague Bill Lowrie spent several years studying the Gubbio formation, sampling up from the Tertiary and down through the Cretaceous. They were first interested in trying to decipher geological history. Their strategy was to correlate reversals in the earth's magnetic field, which left a distinct mark on the grain patterns in rocks with forams in the fossil record. They learned to

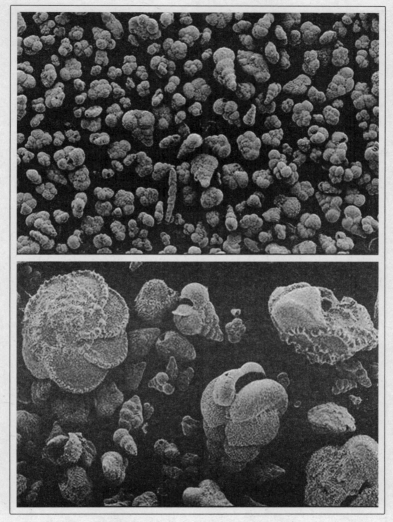

Figure 8.2 Foraminifera of the Tertiary (top) and Cretaceous (bottom). Walter Alvarez was puzzled by the rapid, dramatic change in foram size between the end of the Cretaceous and the beginning of the Tertiary periods, which is seen worldwide. These specimens are from a different location (not Gubbio). Images courtesy of Brian Huber, © Smithsonian Museum of Natural History.

determine where they were in the rock formation by the forams characteristic of certain deposits and by recognizing the boundary between the Cretaceous and Tertiary rocks. That boundary was always right where the dramatic reduction in foram size occurred. The rocks below were Cretaceous, the rocks above were Tertiary, and the thin layer of clay was in the gap between (Figure 8.4). The boundary is referred to as the K-T boundary (K is the traditional abbreviation for the Cretaceous, T, for the Tertiary).

When another geologist, Al Fischer, pointed out that the K-T boundary was about the same age as the most famous extinction of all—that of the dinosaurs—Walter became even more interested in those little forams and the K-T boundary.

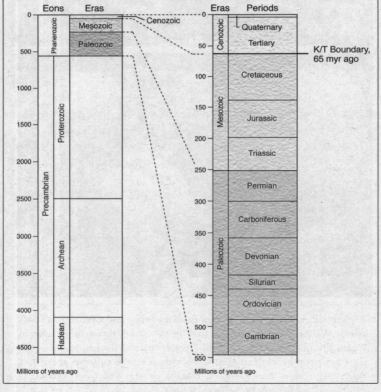

Figure 8.3 Geologic Time Scale. Figure by Leanne Olds.

Walter was relatively new to academic geology. After he had received his Ph.D., he worked for the exploration arm of a multinational oil company in Libya until Colonel Qaddafi expelled all of the Americans from the country. His work on magnetic reversals had gone well, but he realized that the abrupt change in the Gubbio forams and the K-T extinction presented a much bigger mystery.

One of the first questions Walter wanted to answer was: How long did it take for that thin clay layer to form? To answer this he would need some help. It is common for children to get help from their parents with their science projects. However, it is extremely unusual when the "child" is in his late thirties. But few children had a dad like Walter's.

From A-bombs to cosmic rays

Walter's father, Luis Alvarez, knew very little about geology or paleontology, but he knew a lot about physics. A central figure in the birth and growth of nuclear physics, he had received his Ph.D. in physics in 1936 from the University of Chicago and worked at the University of California at Berkeley under Ernest Lawrence, the recipient of the 1939 Nobel Prize in Physics for his invention of the cyclotron.

Luis' early work in physics was interrupted by World War II. During the first years of the war, he worked on the development of radar and systems that would help airplanes land safely with poor visibility. He received the Collier Trophy, the highest honor in aviation, for developing the Ground Controlled Approach system (GCA) for landings in bad weather.

In the middle of the war, he was recruited by the Manhattan Project, the top-secret national effort to develop atomic weapons. Luis and his student Lawrence Johnston designed the detonators for the explosives that would be used in two bombs. Robert Oppenheimer, the director of the Manhattan Project, then put him in charge of measuring the energy released by the bombs. Luis was one of the very few to witness the first two atomic blasts. He flew as a scientific eyewitness to the first test of the atomic bomb, in the New Mexico desert, and then, shortly thereafter, to the bomb dropped on Hiroshima, Japan.

After the war, Luis returned to physics research. He developed the use of large liquid hydrogen bubble chambers for tracking the be-

Figure 8.4 The K/T Boundary at Gubbio. The older, white, fossil-rich Cretaceous limestone below is separated from the darker Tertiary limestone above by a thin layer of clay (marked with a coin) that documents the abrupt change in ocean life at the end of the Cretaceous. Photo courtesy of Frank Schönian, Museum of Natural History, Berlin.

havior of particles. In 1968, Luis received the Nobel Prize in Physics for his work in particle physics.

That would seem to be a nice capstone to an illustrious career. But several years later his son Walter moved to Berkeley to join the geology department. This gave father and son the chance to talk often about science. One day, Walter gave his dad a small polished cross section of Gubbio K-T boundary rock and explained the mystery within it. His father, then in his late sixties, was hooked, and the two started brainstorming about how to measure the rate deposition around the K-T boundary. They needed some kind of atomic timekeeper.

Luis, one of the world's experts on radioactivity and decay, first suggested that they measure the abundance of beryllium-10 in the

K-T clay. This isotope is constantly created in the atmosphere by the action of cosmic rays on oxygen. The more time it took to deposit the clay, the more beryllium-10 would be present. Luis put Walter in touch with a physicist who knew how to do the measurements. But just as Walter was set to begin, he learned that the published half-life of Be-10 was wrong; the actual half-life was shorter, and too little Be-10 would be left to measure after 65 million years.

Luis came up with another idea.

Space dust

Luis remembered that meteorites are ten thousand times richer in elements from the platinum group (platinum, rhodium, palladium, osmium, iridium, and ruthenium) than is the earth's crust. He figured that the rain of dust from outer space should be falling, on average, at a constant rate. Therefore, by measuring the amount of space dust (platinum elements) in rock samples, one could calculate how long they had taken to form.

These elements are not abundant, but they can be measured. Walter figured that if the clay bed had been deposited over a few thousand years, it would contain a detectable amount of platinum group material, but if it had been deposited more quickly, it would be free of these elements.

Luis decided that iridium, not platinum itself, was the best element to measure because it was more easily detected. He also knew who could do the measurements, the two nuclear chemists Frank Asaro and Helen Michel at the Berkeley Radiation Laboratory.

Walter gave Asaro a set of samples from across the Gubbio K-T boundary, but for months he heard nothing. Asaro's analytical techniques were slow, his equipment was not working, and he had other projects to follow.

Nine months later, Walter got a call from his dad. Asaro had results. They had expected iridium levels on the order of 0.1 parts per billion of sample. Asaro found 3 parts per billion in the clay layer, about *thirty times* more than expected and more than the level found in other layers of the rock bed.

Why did that thin layer have so much iridium?

Before they got carried away with speculation, it was important to know if the high level of iridium was an anomaly of rocks around

Gubbio or a more widespread phenomenon. Walter searched for another exposed K-T boundary site they could sample. He found such a place at Stevns Klint, south of Copenhagen, Denmark. He could see right off that "something unpleasant had happened to the Danish sea bottom" when the clay was deposited. The cliff face was almost entirely made of white chalk, full of all kinds of fossils. But the thin K-T clay bed was black, stunk of sulfur, and had only fish bones in it. Walter deduced that when this "fish clay" was deposited, the sea was an oxygen-starved graveyard. He collected samples and delivered them to Frank Asaro.

In the Danish fish clay, the iridium levels were 160 times the background levels.

Something very unusual, and very bad, had happened at the K-T boundary. The forams, the clay, the iridium, the dinosaurs, were all signs—but of what?

It came from outer space

The Alvarezes concluded right away that the iridium must have been of extraterrestrial origin. They thought of a supernova, the explosion of a star that could shower earth with its elemental guts. The idea had been kicked around before in paleontological and astrophysics circles.

Luis knew that heavy elements are produced in stellar explosions so, if that idea was right, there would be unusual amounts of other elements as well in the boundary clay. The most important isotope to measure was plutonium 244, with a half-life of 75 million years. It would still be present in the clay layer but decayed in ordinary earth rocks. Rigorous testing found no elevated level of plutonium. It was disappointing, but at least they had ruled out one theory.

Luis kept imagining scenarios that could account for a worldwide dieoff. He thought that maybe the solar system had passed through a gas cloud, that the sun had become a nova, or that the iridium could have come from Jupiter. None of these ideas held up. An astronomer at Berkeley, Chris McKee, suggested that an asteroid could have hit the earth. At first Luis thought that would only create a tidal wave, and he could not see how even a giant tidal wave could kill the dinosaurs in Montana and Mongolia.

Then he remembered the volcanic explosion on the island of Krakatoa in 1883, a cataclysm that had been well documented. He recalled that miles of rock had been blasted into the atmosphere and that fine dust particles had circled the globe and stayed aloft for two years or more. He also knew from nuclear bomb tests that radioactive material mixed rapidly between hemispheres. Maybe a huge amount of dust from a large impact could have turned day into night for a few years, cooling the planet and shutting down photosynthesis?

If so, how big an asteroid would it have been?

From the iridium measurements in the clay, the concentration of iridium in so-called chondritic meteorites, and the surface area of the earth, Luis calculated the mass of the asteroid to be about 300 billion metric tons. He then was able to infer that the asteroid would have had a diameter of between 6 and 14 kilometers.

That diameter might not seem enormous compared to the 13,000-kilometer diameter of the earth. But consider the energy of the impact. Such an asteroid would enter the atmosphere at about 25 kilometers per second—over 50,000 miles per hour. It would punch a hole in the atmosphere 10 kilometers across and hit the planet with the energy of 10^8 megatons of TNT. (The largest atomic bomb ever exploded released the equivalent of about one megaton—the asteroid was 100 hundred million times more powerful.) With that energy, the crater's impact would be about 200 kilometers across and 40 kilometers deep, and immense amounts of material would be ejected into the atmosphere.

The team had their foram- and dinosaur-killing scenario.

Hell on Earth

The asteroid crossed the atmosphere in about one second, heating the air in front of it to several times the temperature of the sun. On impact, the asteroid vaporized, an enormous fireball erupted into space, and rock particles were launched as far as halfway to the moon. Huge shock waves passed through the bedrock, then curved back up to the surface and shot melted blobs of bedrock to the edge of the atmosphere and beyond. A second fireball erupted from the pressure on the shocked limestone bedrock. For a radius of a few hundred kilometers from ground zero, life was annihilated. Farther away, matter ejected into space fell back to the earth at high speeds—like trillions of meteors—heated up on reentry, and ignited forest fires across continents.

Tsunamis, landslides, and earthquakes further ripped apart landscapes closer to the impact.

Elsewhere in the world, death came a bit more slowly.

The debris and soot in the atmosphere blocked the sun, and the darkness may have lasted for months. This shut down photosynthesis and halted food chains at their base. Animals at successively higher levels of the food chain also succumbed. The K-T boundary marks more than the end of the dinosaurs; it is also the end of belemnites, ammonites, and marine reptiles. Paleontologists estimate that 50 percent of all marine genera, and perhaps 80–90 percent of all marine species, went extinct. On land, nothing larger than 25 kilograms in body size survived.

It was the end of the Mesozoic world.

Where is the hole?

Luis and Walter Alvarez, Frank Asaro, and Helen Michel put together the whole story—the Gubbio forams, the iridium anomaly, the asteroid theory, the killing scenario—in a single paper published in the journal *Science* in June 1980. It is a remarkable, bold synthesis across different scientific fields, perhaps unmatched in scope by any other single paper in the modern scientific literature.

They were concerned that the scientific community was not prepared to accept their theory. The team did as much as possible to test their case, even performing an iridium analysis on a New Zealand K-T bed just to double-check their findings. The sample showed a twenty-fold iridium spike, confirming that the phenomenon was global.

They had good reason to be worried. For the previous 150 years, since the beginning of modern geology, the emphasis had been on the power of gradual change. The science of geology had supplanted the biblical stories of catastrophes. The idea of a catastrophic event on the earth was not just disturbing, it was considered unscientific. Until the asteroid paper, explanations for the disappearance of the dinosaurs usually invoked gradual changes in climate or in the food chain to which the animals could not adapt.

Some geologists scoffed at the catastrophe scenario. Some paleontologists were not at all persuaded by the asteroid theory, pointing out that the highest dinosaur bone in the fossil record at the time was three meters below the K–T boundary. Perhaps the dinosaurs were already gone when the asteroid hit? Other paleontologists pointed out, however, that since dinosaur bones were so scarce, one should not expect to find them right up against the boundary. Rather, they argued, the rich fossil record of forams and other creatures is the more revealing, and forams and ammonites persisted right up to the K–T boundary.

Of course, a larger problem also begged explanation: Where on Earth was that huge crater? To both skeptics and proponents, this was an obvious weakness of the theory, so the hunt was on to find the impact zone if it existed.

At the time, there were only three known craters on the earth 100 kilometers or more in size. None was the right age. If the asteroid had hit the ocean, which covers more than two thirds of the planet's surface, the searchers might be out of luck. The deep ocean was not well mapped, and a substantial part of the pre-Tertiary ocean floor has been swallowed up into the deep earth in the continual movement of tectonic plates.

In the decade after the asteroid theory was proposed, many clues and trails were pursued, often to dead ends. As the failures mounted, Walter began to believe that the impact had in fact been in an ocean.

Then a promising clue emerged from a riverbed in Texas. The Brazos River empties into the Gulf of Mexico, and its sandy bed is right at the K–T boundary. When examined closely by geologists familiar with the pattern of deposits left by tsunamis, the sandy bed was found to have features that could be accounted for only by a giant tsunami, perhaps 50 to 100 meters high.

Many scientists were on the hunt for the site of the impact. Alan Hildebrand, a graduate student at the University of Arizona, was one of the most tenacious. He concluded that the Brazos River tsunami bed was a crucial hint to the crater's location—that it was in the Gulf of Mexico or the Caribbean. Looking at all the maps in search of a candidate crater, he found some rounded features on maps of the sea floor north of Colombia. He also learned of some circular gravity anomalies, places where the concentration of mass varies, on the coast of Mexico's Yucatán Peninsula.

Hildebrand searched for any other hints that he was on the right track. Two signatures of impact events are glass spherules, called tektites, which form under the tremendous temperatures and pressure

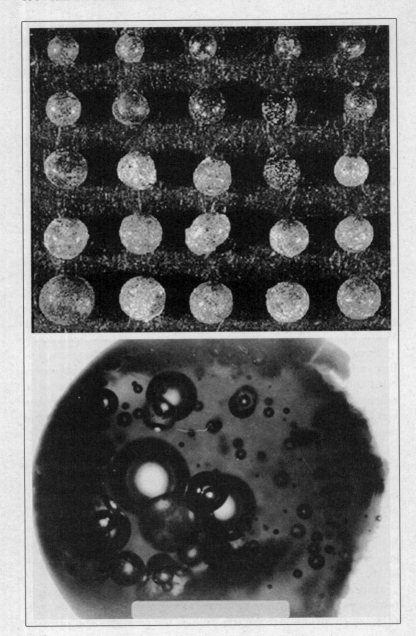

of impact (Figure 8.5), and microscopic "shocked" quartz grains. Hildebrand noticed a report of tektites in late Cretaceous rocks from a site on Haiti. When he visited the lab that had made the report, he recognized the material as impact tektites. He then went to Haiti and discovered that the deposits there included the largest tektites and shocked quartz grains ever found. He and his advisor, William Boynton, surmised that the impact site was within 1,000 kilometers of Haiti.

After Hildebrand and Boynton presented their findings at a conference, they were contacted by Carlos Byars, a reporter for the *Houston Chronicle*. Byars told Hildebrand that geologists working for PEMEX, the state-owned Mexican oil company, might have discovered the crater many years earlier, for Glen Penfield and Antonio Camargo had studied the circular gravity anomalies in the Yucatán. PEMEX would not allow them to release company data, but they did suggest at a conference in 1981, just a year after the Alvarezes' asteroid proposal, that the feature they had mapped might be the crater. Penfield even wrote to Walter Alvarez with that idea.

In 1991, Hildebrand, Boynton, Penfield, Camargo, and their colleagues formally proposed that the crater 180 kilometers in diameter (almost exactly the size predicted by the Alvarez team) one half mile below the village of Chicxulub [Cheech-zhoo-loob] on the Yucatán Peninsula was the long-sought K-T impact crater (Figure 8.6).

Crucial tests were still necessary to determine if Chicxulub was truly the "smoking gun." The first issue was the age of the rock—no easy task to determine because the crater was buried. The best approach would be to test the core rock samples from the wells drilled by PEMEX decades earlier. The results were spectacular. One lab obtained a figure of 64.98 million years, another a value of 65.2 million years. Right on the button—the melted rock was the same age as the K-T boundary.

Figure 8.5 Tektites. Tektites from Dogie Creek, Wyoming (top), and Beloc, Haiti (bottom). These small spheres of glassy rock were formed under the intense heat of impact and rained down across a vast area of the planet. Note the bubbles within the glassy sphere (bottom) that formed in the vacuum of space as the particles were ejected out of the atmosphere. Top photo courtesy of Alan Hildebrand and the Geological Society of Canada. Bottom figure from J. Smit, *Ann. Rev. Earth Planet. Sci.* (1999) 27: 75–113, used with permission.

The Haitian tektites were also dated to this age, as was a deposit of material ejected from the impact. Detailed chemical analyses showed that the Chicxulub melt rock contained high levels of iridium and that it and the Haitian tektites came from the same source. Furthermore, the Haitian tektites had an extremely low water content and the gas pressure inside was nearly zero, indicating that the glass had solidified while in ballistic flight outside the atmosphere.

Within a little more than a decade, what had at first seemed to be a radical and, to some, an outlandish idea had been supported by all sorts of indirect evidence, then ultimately confirmed by direct evidence. Geologists subsequently identified ejected material that covers most of the Yucatán and is deposited at more than 100 K-T boundary sites around the world (Figures 8.7 and 8.8).

The identification of the huge crater, while a great advance for the asteroid theory, was bittersweet for Walter. Luis Alvarez had passed away in 1988, just before its discovery.

Out of the ashes

Luis' ideas, however, have continued to influence geology and the study of the K-T boundary. In 2001, a team of scientists used Luis' idea of measuring the accumulation of space dust particles to estimate the length of the K-T boundary period, the problem that Walter had first asked his dad about. They used an isotope of helium [^{3}He], instead of beryllium or iridium, as a timekeeper. The researchers estimated that the K-T-boundary clay had been deposited in roughly 10,000 years. They also examined one particularly well resolved K-T boundary site in Tunisia, where a very thin (2–3 mm thick) layer is present at the base of the boundary clay. This layer contains shocked quartz grains and other residue of the material ejected from the impact zone that spread throughout the atmosphere and then settled back to earth. This thin fallout layer was estimated to be deposited over a period of about 60 years.

These findings suggest that it took about 10,000 years for ocean food chains and ecosystems to recover to the point where the microscopic fauna were repopulated to the levels present before the impact. But many of the larger ocean and land animals never recovered.

Instead, in the wake of the destruction of the Mesozoic world, a new age emerged—the Cenozoic, the age of mammals. This group of

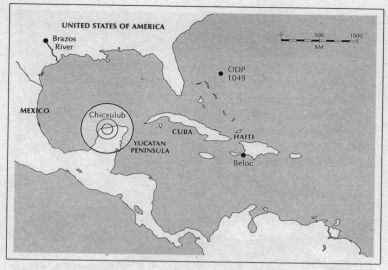

Figure 8.6 Location of Chicxulub crater and key impact evidence sites. The map shows the locations of various impact evidence—the tsunami bed in the Brazos River, tektites in Haiti, the ocean drilling site 1049 (see Figure 8.8), and the crater and surrounding ejected material on the Yucatán Peninsula. Map by Leanne Olds.

generally small species, such as those found on Andrews' Asiatic expedition, took advantage of the niches vacated in the demise of the Cretaceous. Mammals evolved rapidly into species of many sizes, including large herbivores and carnivores. Within 10 million years, forms representing most modern orders appear in the Cenozoic fossil record. The dinosaurs' demise was certainly the mammals' gain.

Other smoking guns?

The lessons learned from Chicxulub have spawned intense interest in identifying other impact events that may have affected life on Earth. The K-T extinction was not the largest on record. That dubious honor belongs to the Permian-Triassic transition, when perhaps as many as 90 percent of species living in the late Permian disappeared in less than 200,000 years about 251 million years ago. While many ideas for this mass extinction are being investigated, two

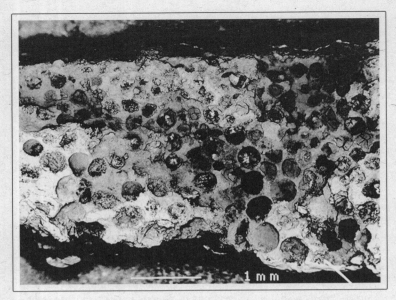

Figure 8.7 A K/T Spherule Layer. An exquisitely well preserved site near Tbilisi, Republic of Georgia. A magnified view of the deposit reveals a graded layer of spherules (smaller particles at the top, larger at the bottom) ejected from the impact that is also highly enriched in iridium (86 ppb). Image from J. Smit, *Ann. Rev. Earth Planet. Sci.* (1999) 27: 75–113, used with permission.

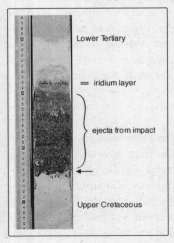

Figure 8.8 The K/T impact documented in an ocean core sample. This core sample, drilled at a site about 500 km east of Florida (Ocean Drilling Project site 1049), beautifully depicts the K/T event. The scale at the left shows that a nearly 15-cm-deep layer of ejected material was deposited on the ocean bottom hundreds of miles from the impact site. Note the iridium-containing layer that settled on top of the denser ejected material. Image courtesy of Integrated Ocean Drilling Program.

large candidate craters were recently proposed as evidence of a Permian impact, the 200-kilometer Bedout crater off the northwest coast of Australia and an even larger crater buried under the ice of Wilkes Land, Antarctica.

The Chicxulub discovery has also inspired astronomers to scan the sky for other potential inbound asteroids. There are thousands of asteroids close to the earth's orbit. On March 23, 1989, a 1,000-foot-diameter asteroid missed the earth by 400,000 miles, passing through the exact position that the planet had occupied just six hours earlier. Had it struck, it would have caused an explosion greater than 1,000 megatons, the largest in recorded history.

We now understand that the history of life on Earth has not been the orderly, gradual process envisioned by generations of geologists since Lyell and Darwin. There are more than 170 confirmed impact sites of various sizes on our planet—and more to come.

We also now understand that while the K-T extinction has long been said to mark the end of the dinosaurs, that is not true. As we will discover in the next chapter, one group is still very much alive.

Figure 9.1 **Dinosaur tracks.** When tracks like these were first found in the 1800s, they were thought to be ancient bird tracks. This Cretaceous trackway includes footprints of several species and is exposed on "Dinosaur Ridge" outside Morrison, Colorado. Photo © Joe McDaniel, used by permission.

Dinosaurs of a Feather

The bird is not in its ounces and inches,
but in its relations to Nature.
— *Ralph Waldo Emerson*

IT MAY SEEM HARD to believe, but there was a time when paleontologists were not that interested in finding dinosaurs. By the 1930s, after the dinosaur rushes of the previous decades, museum halls and storerooms were packed with the giant beasts, so they were no longer prized finds. Besides, they were generally viewed as sluggish, lumbering reptiles that had gone extinct, perhaps merely as victims of their own ponderousness. Researchers moved on to other quarry that had modern descendants, such as mammals.

But not John Ostrom. His fascination with dinosaurs grew out of an early interest in evolution. In the late 1940s, as a premedical student at Union College in his hometown of Schenectady, New York, he was required to take a course on evolution. The night before the course started, he began to read *The Meaning of Evolution,* by the paleontologist George Gaylord Simpson. Ostrom was hooked. He stayed up all night finishing it and even wrote to Simpson to say how the book had gripped him. In turn, Simpson invited Ostrom to come study with him at Columbia. Much to the dismay of his father, a physician, Ostrom abandoned the premed path and switched his major to geology (sound familiar?).

In 1951, Ostrom went to Columbia to pursue his Ph.D., but he decided not to study fossil mammals with Simpson but duck-billed dinosaurs with Ned Colbert, the lone and very passionate curator of reptiles at the American Museum of Natural History. Later, after teaching stints at Brooklyn College and Beloit College, Ostrom became curator of vertebrate paleontology at Yale's Peabody Museum. He was responsible for an enormous collection of dinosaurs gathered in the American West in the late nineteenth century by the legendary O. C. Marsh.

Ostrom immediately began his own field expeditions to the Cloverly Formation of Montana and Wyoming. Starting in the summer of 1962, his team explored the colorful badlands. By the end of the third season, they had found several dozen promising sites. Late one August afternoon in 1964, Ostrom and his assistant Grant Meyer were marking and plotting each site for future excavation. Walking between two sites, they spotted several claws and bones just a few yards away, down a slope to their right. They both nearly tumbled as they raced to the spot where a large clawed hand was sticking up out of the dirt.

As they were only marking sites that day, they did not have the picks, chisels, shovels, and other necessary tools to dig up fossils. Using just a jackknife, a small paintbrush, and a whiskbroom, they quickly uncovered more of the hand, some ribs, a vertebra, and a complete foot before darkness approached.

All that night, Ostrom thought about the bones. He was sure they belonged to a small carnivorous dinosaur, but what kind he could not say: "I was almost certain, although still wary, that we had discovered something totally new."

The next day and each day for the following week he and Meyer returned to the site and continued removing the rock and loose dirt from around the bones, then packed them for transport. After eight days of slow, careful work, they had parts of two skeletons.

Only about three and a half feet tall and probably weighing about 150 pounds in the flesh, the creature was small by dinosaur standards. But the foot had never been seen before. Unlike other carnivorous dinosaurs, which had three symmetrically arranged main toes and one smaller toe on the inner side of the foot, this creature's outside toe was as long as its middle toe. But its most remarkable feature was the second toe, which was quite long and carried a huge, retractable, sickle-like claw (Figure 9.2). Ostrom realized instantly that this appendage was not for digging or climbing but for cutting and slashing prey. He would dub the animal *Deinonychus* ("terrible claw").

Ostrom suspected his discovery was something new and interesting. Little did he know, however, as he prepared for the long drive home, that it would come to be seen as the most important dinosaur discovered in the twentieth century. *Deinonychus* and John Ostrom would change our pictures of dinosaurs forever: they weren't lumbering oafs, they weren't stupid, they weren't cold-blooded, and, most startling of all, *they weren't extinct.*

To understand how Ostrom overturned established conceptions and the importance of his ideas in the bigger picture of the history of life, we have to return to an epic debate involving dinosaurs and the most famous and important fossil discovered in the nineteenth century—a fossil that took center stage in the first debates over Darwin's new theory of evolution and that would also play a pivotal role in Ostrom's resurrection of the dinosaurs.

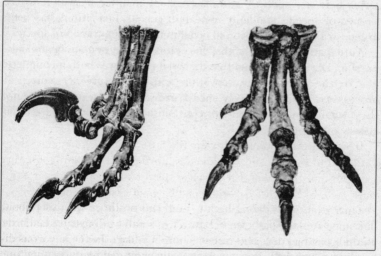

Figure 9.2 The "Terrible Claw." Left, the unusual asymmetrical foot of *Deinonychus* with its large retractable claw on the second toe compared with, right, the more typical symmetrical foot of the theropod *Allosaurus*. Photos from J. Ostrom (1969), *Discovery* 5 (1): 1–9, courtesy of Yale's Peabody Museum of Natural History.

Missing links

When Darwin finally put forth his theory of evolution after more than twenty years of gathering evidence from geology, zoology, botany, animal breeding, paleontology, and biogeography, he was still lacking (by his own admission) some critical pieces of the puzzle. Perhaps the most obvious, and the most troubling in terms of gaining acceptance, was the absence of transitional forms — species that connected one group of organisms to another.

Everyone knew that the living world was made up of very distinct groups such as fish, reptiles, mammals, and birds. If, as Darwin proposed, all of these animals were descended from a common ancestor and gradually modified, why did such great gaps exist between them?

Darwin asked aloud: "As by this theory innumerable transitional forms must have existed, why do we not find them embedded in countless numbers in the crust of the earth?" He was painfully honest that "geology assuredly does not reveal any such finely graduated series between species" and "the abrupt manner in which whole groups of species suddenly appear in certain formations has been urged . . . as a fatal objection to belief in the transmutation of species."

Anticipating, correctly, that his critics would pounce on this admission, Darwin explained that the fossil record was both incomplete, due to the rarity of preservation and the nature of geological processes, and also relatively unexplored. A crucial test of his theory, then, would be the discovery, or continuing absence, of links between classes of organisms.

It was, mercifully, a short wait.

The first bird

For many centuries dating back to even Roman times, the exceptional limestone around Solnhofen in Bavaria, in southern Germany, had been used for paving roads and constructing buildings. In the late 1700s, a budding playwright, Johann Alois Senefelder, began experimenting with using these stones for printing, rather than copper plates. He discovered a new way of reproducing images by impregnating slabs with ink and etching them with acids. He called it "stone-printing," but it was the French-based term "lithography" that caught on, and certain artists such as Eugène Delacroix mastered the new form.

As the demand for lithographs grew, the quarrying operations at Solnhofen expanded. Any blemish on these very fine grained slabs would mar the process, so each slab was carefully inspected. Since it had been deposited in a shallow sea 150 million years ago during the late Jurassic period, the rock also harbored various fossils from that time—such as shrimps, crabs, insects, and fish—that were often exceptionally well preserved. While the fossils ruined the slabs for lithography, they were a boon to scientists such as Hermann von Meyer.

The author of a five-volume work, *Fauna of the Ancient World*, von Meyer was one of Germany's most respected paleontologists. He had identified many of the creatures found at Solnhofen, including a number of pterosaurs (an extinct group of flying reptiles). But in late 1860 or early 1861, the workers found a slab with the impression of a feather on it. Von Meyer was worried at first that it was some kind of hoax, for no bird had ever been found in Mesozoic rocks, and the group was thought to be considerably younger. But he decided the fossil was genuine and published a short description of it in a German journal; he cautioned that since there was no skeleton, the feather did not necessarily belong to a bird.

No sooner had this report appeared that in the very next month, September 1861, less than two years after the publication of *On the Origin of Species*, von Meyer announced a much more stunning find— a nearly complete skeleton with impressions of feathers surrounding its forelimbs and long bony tail. He christened it *Archaeopteryx lithographica*, with *Archaeo* meaning "ancient," *pteryx* meaning "winged," and the species name reflecting the remarkable stone in which it had been preserved and discovered (Figure 9.3).

Word of the ancient winged creature of Solnhofen spread quickly. The specimen was in private hands, and the owner wanted to sell it. The anatomist Richard Owen, the superintendent of the British Museum, though openly and vehemently opposed to Darwin's theory, was nevertheless eager to secure the fossil for Britain. He persuaded the museum's trustees to authorize the then considerable sum of 700 British pounds to purchase the treasure.

Dirty Dick and Darwin's Bulldog

In procuring the specimen, Owen also ensured that he would be the first British scientist to see it. An expert anatomist, he was also

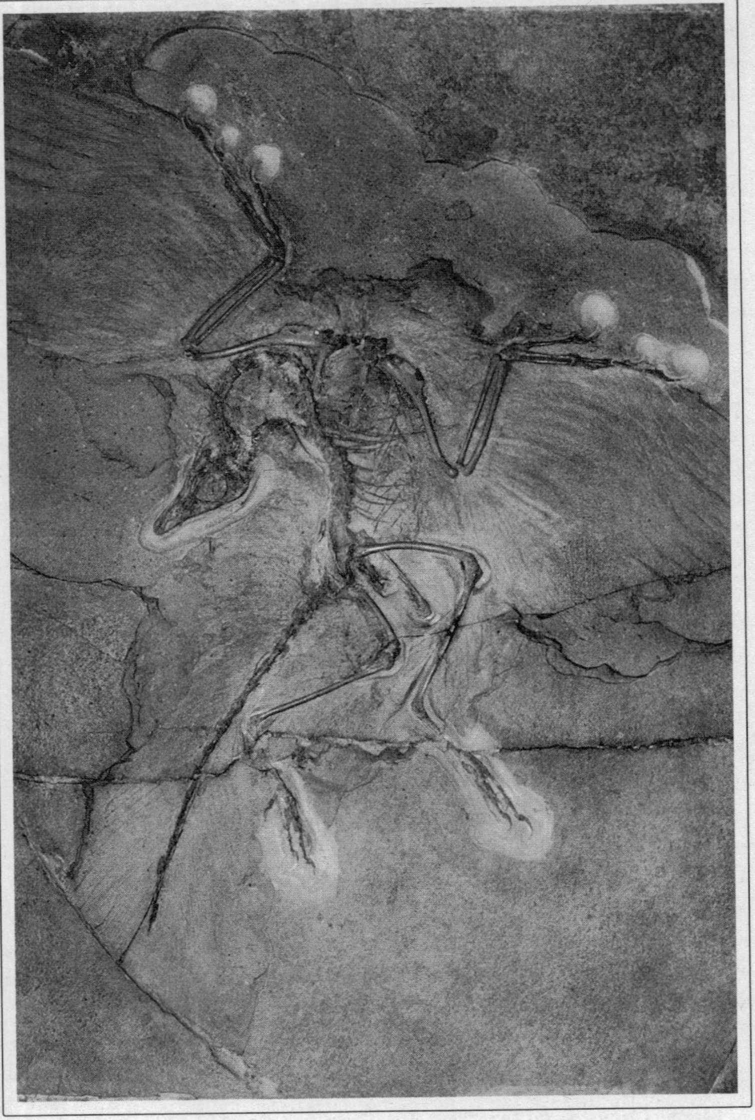

Figure 9.3 *Archaeopteryx*. The Berlin specimen is shown; details are easier to see in it than in the original 1861 specimen. Photo courtesy of Luis Chiappe.

acutely concerned about the role the creature might play in the heated debate over Darwin's new theory. It was a curious beast. The body had both reptile-like and bird-like features (the head was not uncovered until later). The reptilian characteristics included a bony tail, three clawed fingers, and aspects of its ribs and vertebrae, while its feathers were certainly like those of a bird. The feathers and the creature's small size apparently decided the matter for Owen. In November 1862, he read an account before the Royal Society of London in which he compared *Archaeopteryx* with other winged creatures, pterosaurs and birds, and declared it "unequivocally to be a Bird."

Darwin, saddled by an outbreak of one of his illnesses, missed the chance to see the fossil for himself or to hear Owen's interpretation. He first learned of the creature from his trusted friend, the expert paleontologist Hugh Falconer:

> You were never more missed—at any rate by me—for there has been this grand *Darwinian* case of the *Archaeopteryx* for you and me to have a long jaw about. Had the Solenhofen quarries been commissioned—by august command—to turn out a strange being à la Darwin—it could not have executed the behest more handsomely—than in the *Archaeopteryx*.

Falconer went on to skewer Owen's account:

> You are not to put your faith in the slip-shod and hasty account of it given to the Royal Society. It is a much more astounding creature —than has entered into the conception of the describer.

Falconer saw *Archaeopteryx* as not simply a "bird" but as a "sort of misbegotten-bird-creature—the dawn of an oncoming conception à la Darwin"—an ancestor of birds, a missing link.

Darwin quickly wrote back to Falconer asking for more details and wrote to others describing the fossil bird as "a grand case for me; as no group was so isolated as the Birds; & it shows how little we know what lived during former times."

Owen had his reasons, perhaps twenty years' worth, for minimizing *Archaeopteryx*. It was he who, in 1842, coined the name *Dinosauria* for the massive extinct fossil reptiles that were being dug up around Britain and Europe. But it was also he who, in the very same publication, used the existence of dinosaurs to rebuff the early concepts of evolution then circulating.

Various naturalists, including Jean-Baptiste Lamarck and Étienne Geoffrey St.-Hilaire, had tried to explain the gradual appearance of animals in the fossil record. The succession of fish, reptiles, mammals, and birds was seen as evidence of a progressive process, with the later forms as "higher," improvements over earlier forms. Owen wanted to refute this idea. Progressive evolution meant that modern reptiles should be more advanced than extinct ones. Yet he saw the era of the dinosaurs as the pinnacle for reptiles: "The period when the class of Reptiles flourished . . . in the greatest number and the highest grade of organization, is past; and since the extinction of the Dinosaur order, it has been declining." For him, there was no progress and, hence, no evolution.

Owen argued instead that "the different species of Reptiles were suddenly introduced upon the earth's surface" and that the characteristics that define dinosaurs "were impressed upon them at their creation, and have been neither derived from improvement of a lower, nor lost by progressive development into a higher type." He thought the same would be true of birds, and their abrupt appearance in the Tertiary fossil record served to confirm that until *Archaeopteryx* appeared.

Owen was highly influential in the British scientific establishment. In any matter touching on evolution, especially those involving comparative anatomy, he led the opposition, generally by discounting inconvenient observations and concocting others. He also used his political connections to maneuver himself into important positions to the exclusion of other scientists. Falconer, like Darwin and Thomas Huxley, had squabbled professionally with Owen and witnessed his underhanded tactics; he dubbed him "Dirty Dick." By the early 1860s, Owen had made quite a few enemies, the most formidable of whom was Huxley, a.k.a. "Darwin's Bulldog," who savored any chance to knock Owen off his pedestal. *Archaeopteryx* presented a golden opportunity, and Huxley was more than up to the task.

He approached the matter as one at the crux of the debate over the reality of evolution:

It is admitted on all sides that existing animals and plants are marked out by natural intervals into sundry very distinct groups . . . and out of this fact arises the very pertinent objection,—How is it, if all animals have proceeded by gradual modification from a common stock, that these great gaps exist?

We, who believe in Evolution, reply, that these gaps were once non-existent; that the connecting forms existed in previous epochs of the world's history, but that they have died out.

Naturally enough, then, we are asked to produce these extinct forms of life.

Huxley explained that such evidence is like the deeds of an estate, and that just as an owner must be able to produce such deeds, so too must those who hold to the theory of evolution. He offered that while he could not show the complete document, he was now able to show a considerable "piece of parchment" belonging to it.

Focusing on the differences between reptiles and birds, Huxley posed two questions:

1. Are any fossil Birds more reptilian than any of those now living?
2. Are any fossil Reptiles more bird-like than living reptiles?

His main exhibit in replying yes to the first question was *Archaeopteryx*. Summarizing the anatomical facts—the presence of clawed, separate fingers and the long, bony tail—Huxley concluded, "Thus it is a matter of fact, that in certain particulars, the oldest known bird does exhibit a closer approximation to reptilian structure than any modern bird."

Turning to the question of bird-like reptiles, Huxley first examined the features of pterosaurs and duly rejected them as a connection between reptiles and birds. Then he turned to the dinosaurs and remarked how their hind limbs closely resembled those of birds. But the best link, Huxley saw, was reflected in a single dinosaur, *Campsognathus longpipes*, also recently discovered in the Solnhofen limestone. This small animal was clearly both a dinosaur (by the very criteria Owen used to define the group) but also much more like a bird in anatomy and posture than any dinosaur previously described.

Huxley went on to point out the similarity between developing birds and dinosaurs and between the fossilized tracks that had been found in rocks and those of modern birds (Figure 9.1). "Surely, there is nothing very wild or illegitimate in the hypothesis that . . . the class Aves [Birds] has its root in Dinosaurian reptiles."

Thus Huxley countered Owen with dinosaurs, Owen's very own conception, to bolster the theory of evolution.

The dinosaur-bird link gathered some significant support and the transitional character of *Archaeopteryx* was further sustained when the skull of the British Museum's specimen was located in the slabs and shown to possess teeth, like a reptile's, but a large brain, like that of a bird.

In the 1860s, however, relatively few dinosaurs were known, and as the number and variety of known species grew over the next several decades, confusion grew over the relationships between reptiles and dinosaurs and dinosaurs' connection to birds. All sorts of evolutionary schemes were put forth. The main issue was whether dinosaurs were birds' immediate ancestors or whether their resemblances reflected their descent from a common but more remote Triassic reptilian ancestor (Figure 9.4). All naturalists generally agreed that birds, crocodilians, dinosaurs, pterosaurs, and some other extinct reptiles were descended from a common "archosaur" ancestor. But the differences found among dinosaurs led many to conclude that birds were not their immediate descendants.

Some scientists who continued to pursue the dinosaur-bird link were intrigued by resemblances between certain types of dinosaurs and birds, so that virtually every major group had been proposed at one time or another as bird ancestors. Gerhard Heileman, in an influential book, *The Origins of Birds* (1926), remarked on the similarities between certain dinosaurs (coelurosaurs) and birds but rejected the connection on the basis of the absence in coelurosaurs of a wishbone or of the collar bones (clavicle) that were thought to form the bird's wishbone. Since this structure was considered one of the essential parts of being a bird, Heileman concluded that the coelurosaurs could not possibly be bird ancestors and the likely reptilian bird ancestor was some much earlier Triassic archosaur.

And so the dinosaur-bird link faded away until John Ostrom started assembling what he had found in the Montana badlands.

Renaissance

The teeth were like steak knives, so *Deinonychus* was definitely a carnivore, and its hands were capable of grasping. These two characteristics defined it as a "theropod" dinosaur, a large group that includes the icon *Tyrannosaurus rex*. But while *Deinonychus* was just

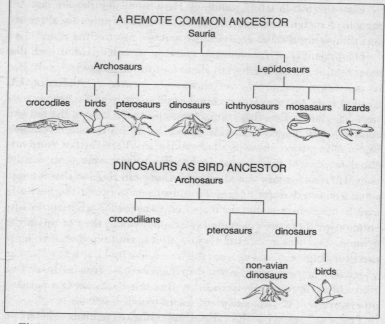

Figure 9.4 **Alternative views of dinosaur and bird relationships.** Top, a long-standing view of birds and dinosaurs was that they represented independent lines of evolution descended from a common archosaur ancestor that was shared with crocodiles and pterosaurs. Bottom, in an alternative view argued by Thomas Huxley and others, birds evolved from dinosaurs. Drawing by Leanne Olds.

one fifth the height of *T. rex*, its foot claw was as large as any *T. rex* claw.

Ostrom tried to picture *Deinonychus* behavior and posture. Mixed in among the *Deinonychus* bones were those of the much larger plant eater *Tenontosaurus*, which Ostrom figured to be the carnivore's dinner. How did it take down larger prey? Its hands could not be used for walking, so it clearly walked on its hind limbs. How then did *Deinonychus* use those massive foot claws? It would have to jump or pounce on its victims. But, Ostrom realized, who ever heard of a reptile with the agility to walk on its hind legs and to have the balance necessary to attack while standing? He wrote: "Reptiles are just not capable of such maneuvers. . . . As we all know, reptiles are sluggish, sprawling animals that are boringly inactive most of the time."

The assembly of *Deinonychus'* skeleton revealed that it had, like most reptiles, a rather long tail about half the length of the body. But its tail was unique, for the vertebrae were encased in bundles of thin parallel rods. Ostrom was puzzled until he realized that these rods were like the tendons that connect muscles to the base of the tail and allow it to be flipped from side to side, as in lizards and crocodiles. But in *Deinonychus*, these tendons ran the length of the tail, were oriented to control the up-and-down motion, and were ossified into bone. He realized this would make the long tail rigid so that it could act as a counterbalance during the animal's movements. The neck and back bones indicated that the main body was held in a horizontal orientation with the neck curving up. The tail would then be up off the ground. This was not a lumbering, tail-dragging reptile but an agile predator (Figure 9.5). Far from the lizard-like mold, it was more like an ostrich or cassowary, *a bird*. Furthermore, he reasoned, such an active lifestyle raised the possibility that this dinosaur, and perhaps others, was not cold-blooded but warm-blooded, like birds.

This was heresy, but the various resemblances to birds led Ostrom to reconsider the then abandoned idea of a direct dinosaur-bird link. To do so, he had to return to the very fossil that started it all, *Archaeopteryx*. Shortly after completing the first full reports on *Deinonychus* in 1969, he headed to the various museums of Europe to examine the evidence for himself.

What's in a label?

By 1970, just four specimens of *Archaeopteryx* were known, the original feather and three skeletons. Such rare fossils were therefore as treasured by natural history museum directors and private collectors as rare works of art. Every museum would like to have an *Archaeopteryx*, but that even a few specimens exist is remarkable when one considers that they were delicate, hollow-boned animals and that, as land-dwelling animals, they would not be buried in shallow seas very often. The preservation of soft tissues, such as feathers, was even less likely.

Ostrom's pilgrimage took him to London, Marburg, Berlin, and the Solnhofen quarries. In addition to the few *Archaeopteryx*, Ostrom studied the skeletons of pterosaurs that had been collected from the Solnhofen fossil beds. At the Teyler Museum in Haarlem, the Netherlands, he was shown a pterosaur specimen that had been collected in 1855.

Right away, however, Ostrom saw that this fossil was not a pterosaur. Tipping the slab into the sunlight, he saw feathers! He could hardly believe his own eyes—it was another *Archaeopteryx*.

The creature had been displayed in Haarlem, unrecognized, for more than 100 years. Ironically, the specimen had been reported and (mis)identified in 1857 by none other than Herman von Meyer, who christened the "first" *Archaeopteryx* in 1861. (I say "first" because the fossil Ostrom rediscovered was clearly found before the 1861 specimen, but it was not recognized as being something new.)

Ostrom showed his host the feather impressions with some trepidation, for he was sure that the treasure would be snatched from his hands. Stunned, his host picked up the fossil slabs and took them away. Ostrom thought to himself, "You blew it, John. You blew it!" He figured that was the last he would see of only the fifth *Archaeopteryx* identified in 120 years. His host returned fifteen minutes later with

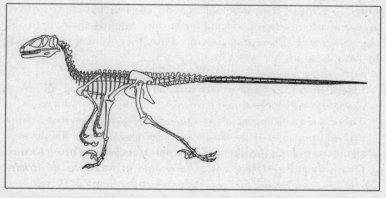

Figure 9.5 *Deinonychus*. A reconstruction of the agile predator's skeleton with the long tail held off the ground for balance. From J. Ostrom (1969), *Discovery* 5 (1): 1–9, courtesy of Yale's Peabody Museum of Natural History.

an old shoebox with a piece of string tied around it. Inside were the slabs. He handed it to Ostrom and said, "Here, here, Professor Ostrom, you have made the Teyler Museum famous." Ostrom flew home with the shoebox on his lap, its contents insured for $1 million.

Birds from dinosaurs, birds as dinosaurs

In order to report his discovery—actually his rediscovery—Ostrom had to examine the new specimen's bones carefully and compare them with what he had found in other *Archaeopteryx*. He said to himself, "Whoa, wait a minute. All of this anatomy—hey, I've seen this before on a larger scale." The more he looked at *Archaeopteryx*, the more dinosaurian features he saw.

The jaw, the teeth, the vertebral column, and parts of the shoulder all resembled aspects of theropod dinosaurs. But the most striking and detailed similarities were between the arms, hands, and wrists of *Archaeopteryx* and the corresponding parts of what he had found in *Deinonychus*. For example, the wrists of both *Archaeopteryx* and *Deinonychus* contained a half-moon-shaped ("semi-lunate") carpal bone, which allowed the wrist to swivel. Critical for allowing the upstroke and downstroke of the wingbeat in flying birds, the bone is found only in some theropods and birds. Ostrom concluded that there were too many of these resemblances for all of them to have evolved independently in theropods and birds. The detailed anatomy of *Archaeopteryx* and *Deinonychus* convinced him they were close relatives. In 1973, he stuck his neck out all the way:

> Indeed, if feather impressions had not been preserved all *Archaeopteryx* specimens would have been identified as . . . dinosaurs. The only reasonable conclusion is that *Archaeopteryx* must have been derived from an early or mid-Jurassic theropod . . . The additional significance of this phylogeny is that "dinosaurs" did not become extinct without descendants.

Ostrom was not just saying that birds evolved from a dinosaurian ancestor but that they were, in fact, dinosaurs—more specifically, theropod dinosaurs.

The very close relationship he asserted was poignantly demonstrated that same year when another "new" specimen of *Archaeopteryx* was reported. A theropod fossil that had been discovered in 1951 and identified as *Compsagnathus* (the same dinosaur pointed to by Huxley as evidence for a dinosaur-bird link) was reexamined and recognized as *Archaeopteryx*. Fifteen years later, the next *Archaeopteryx* identified was again a rediscovery of a purported *Compsognathus* fossil. With three of the six *Archaeopteryx* skeletal specimens originally misidentified as pterosaurs or dinosaurs, Ostrom's assertion, that were it not for feathers, *Archaeopteryx* would have been classified as a dinosaur, rang very true.

But Ostrom's theropod-bird theory was a radical departure from the previous fifty years of thought about bird origins. Some paleontologists were excited and intrigued by Ostrom's view. Other scientists, some ornithologists in particular, were skeptical or dismissive and clung to the previous view of a more remote common ancestor of dinosaurs and birds. Part of the disagreement was driven by emphases on different body characters, with Ostrom and other theropod-bird advocates stressing one set of features to draw a direct theropod-bird connection and their detractors pointing to characters of modern birds or *Archaeopteryx* that they believed indicated a connection to much older reptiles.

Parallel to this revolution in dinosaur-bird relationships, another revolution was unfolding in phylogenetics—the science of evolutionary relationships. A new approach, called "cladistics," aimed for a more objective and quantitative analysis of species relationships. The basic idea was to use characters shared by some but not all taxa to identify sister relationships and to establish evolutionary trees. When Jacques Gauthier, then at the University of California at Berkeley, applied this new approach to reptiles and birds in the mid-1980s, he found strong support that birds are in fact theropod dinosaurs.

This new analysis did not satisfy all of the detractors; at least two of them were quoted as saying that the scientists using cladistics were "full of crap." For more than twenty years some doubters held out, while various supporters were, of course, looking for more evidence. Both were in for a shock.

Which came first, the bird or the feather?

The buzz was all over the meeting halls. The Society for Vertebrate Paleontology was holding its 1996 meeting at the venerable American Museum of Natural History. Many scientists were presenting their latest findings in a broad swath of topics. But the hot new data in this case was limited to a 3-by-5 photograph brought to the meeting by Dr. Philip Currie, a leading dinosaur researcher and head of the Tyrell Museum in Drumheller, Alberta, who had just returned from China. On a visit to Beijing University, he had been shown a new fossil found by a farmer in Liaoning Province. When he looked at the slab, he was "bowled over." It was the fossil of a yard-long dinosaur, something like *Compsognathus*, with a feathery down running along its back. The picture left Ostrom "in a state of shock" and "weak in the knees." Could it be true—a feathered dinosaur?

Currie knew little more about the creature except that he was sure the feathers were not the kind used for flight. But, if the fossil was really as it appeared, it would seem to clinch Ostrom's theory.

Many scientists were eager to learn more about the Chinese specimens. An arrangement was made between the Academy of Natural Sciences in Philadelphia and authorities in China to send a delegation to see the fossils. In the spring of 1997, five scientists, including Ostrom (by then retired), made the journey. Among the fossils they were shown were specimens of the animal known as *Sinosauropteryx*, which caused the sensation at the meeting in New York. Ostrom declared, "This was one of the most exciting moments of my life." He had never expected to see anything like it in his lifetime.

It was just the beginning. A succession of feathered theropod dinosaurs have since been unearthed, some with more advanced, veined feathers than *Sinosauropteryx*, such as the spectacular *Sinornithosaurus millenii* (Figure 9.6). Technically, these dinosaurs are referred to as "non-avian" feathered dinosaurs—they have feathers but do not have wings. The presence of feathers in several theropod groups demonstrates, without a doubt, that feather evolution was well along before the origin of birds and suggests that these structures were adaptations for warmth and insulation before the origin of bird flight. In fact, over the past twenty years, intense research has revealed that other features once thought to be unique to birds, such as their wishbone, swivel-type wrists and even their nesting and egg-laying behaviors, were present in some theropods. As the prominent paleontologist Luis Chiappe put it, "As we discover more and more theropod lineages that acquired birdness, or evolved bird-

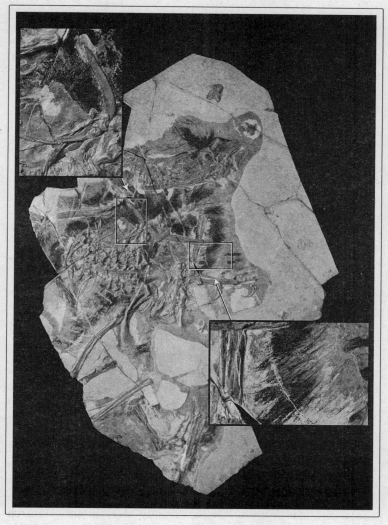

Figure 9.6 A feathered dinosaur. *Sinornithosaurus millenii*, a recently discovered feathered dinosaur from the Yixian Formation near Sihetun, China. Note the prominent wishbone and feathers (boxes), two characteristics of all birds. Photo courtesy of Luis Chiappe.

like features, the line between what is a bird and what is not becomes increasingly blurred."

The blurring of lines is the nature of evolutionary transitions, whether from dinosaurs to birds or, as we will see in the next chapters, from fish to amphibians or between ancient hominids.

Deinonychus *goes to Hollywood*

Ostrom's discoveries and studies of *Deinonychus* and *Archaeopteryx* did more than illuminate the evolutionary transition from dinosaurs to birds and settle a long-standing scientific mystery and debate. They transcended the scientific world and redrew the picture of dinosaurs in the public's imagination. Even before Ostrom received the full scientific acclaim he had so well earned, he had the satisfaction of seeing his vision of dinosaurs come to life in a much larger arena than academic paleontology.

One day in the late 1980s, Ostrom's phone rang. "Professor Ostrom, this is Michael Crichton," the caller said. The author of the *Andromeda Strain* was doing research for a new novel, and he had some questions about the creature Ostrom had discovered in Montana. He wanted to know if it was a meat eater, and whether it could run as quickly as a human or jump as high.

Ostrom told Crichton what *Deinonychus* might do or might be able to do. He also said that *Velociraptor* was its closest relative. Crichton explained apologetically that he had decided not to use the name of Ostrom's animal in his novel, since the Greek name was too difficult. *"Velociraptor"* was more "dramatic." That is how a dinosaur from Mongolia, discovered by Roy Chapman Andrews' team but unknown to the world, wound up being the fearsome, intelligent, and agile predator of Crichton's *Jurassic Park;* they were then brought vividly and terrifyingly to life by the special effects in the blockbuster movie directed by Steven Spielberg.

This one bit of dramatic license aside, Ostrom's discoveries and theory got their cinematic due. In the first scene in which the paleontologist Dr. Alan Grant appears, set in the badlands of Montana, he uncovers the six-inch-long retractable claw of a predator that he deduces to have hunted in packs and dined on *Tenontosaurus*. He then points out to a volunteer group a half-moon-shaped bone in the raptor's wrist and remarks, "No wonder these guys learned to fly." This draws laughter, as most of the group had not read Grant's book on the subject. But young Tim Murphy had read it, and he later asks Grant:

"You really think dinosaurs turned into birds? And that's where all the dinosaurs went?"

Grant replies, "Well, uh, a few species—may have evolved, uh—along those lines—yeah."

Yeah, indeed.

Figure 10.1 **Ellesmere Island. Sor Fiord.** The rocky landscape is exposed only briefly during the short summer. In the foreground, expedition members scout for fossils. Photo courtesy of Neil Shubin.

∽ 10 ∽
It's a Fishapod!

Organic life beneath the shoreless waves
Was born and nurs'd in ocean's pearly caves;
First forms minute, unseen by spheric glass,
Move on the mud, or pierce the watery mass;
These, as successive generations bloom
New powers acquire, and larger limbs assume;
Whence countless groups of vegetation spring,
And breathing realms of fin, and feet, and wing.
—*Erasmus Darwin, "The Temple of Nature" (1802)*

IN THE SUMMER OF 1976, Neil Shubin got caught up, along with the rest of the city of Philadelphia, in the celebration of the American Bicentennial. Having grown up in one of the birthplaces of the Revolution, Neil was surrounded by colonial history. That year he took a strong interest in the archaeology of various ruins around the city. Though just a high school student, Neil got the chance to work under the guidance of a University of Pennsylvania professor on unearthing the history of some old paper mills.

His assignment was to determine what one particular mill produced. Every day, no matter how hot and muggy it was, Neil went to the ruins of the site along the banks of Mill Creek and dug in the dirt for clues. It was fun to find potsherds and old tools, but it was also hard, messy work. Neil figured there had to be other ways to find out about the past than just through physical artifacts, and he went to the library. Scouring microfilms of nineteenth-century trade journals and newspapers, he was able to trace the name of the mill, the cause of the fire that destroyed it, and to discover just what the mill produced. Neil was very proud of his first archaeological success, and he learned three lessons that would serve him well in the years to come: that one could know the past, that he loved fieldwork, and that it paid to do one's homework.

As a student at Columbia University, Neil soon discovered that there were even bigger mysteries lurking in the history of life, and he decided that paleontology, not archaeology, was his calling. He entered graduate school at Harvard, eager to go on expeditions and find fossils. Under the tutelage of Farish Jenkins, his Ph.D. advisor, he got a chance to look for early mammal fossils in sites Jenkins had discovered in Arizona. That experience taught Neil how to spot teeth and bone among a sea of rock, and, as these were often tiny, delicate fossils, he also learned the patience necessary to scour an exposure with a hand-held lens to find and extract individual gems. The search for these buried treasures gave Neil the appetite to find his own fossil sites and to launch his own expeditions.

The mammal fossils he was after were in 200-million-year-old rocks. Lacking the funds for more exotic locales, Neil rented a minivan and drove to nearby Connecticut, where rocks of that age had been known for a long time. But he found zilch. In Connecticut, most of the rock was covered with forest and other greenery; he needed larger exposures than the occasional highway road cut, such as a seacoast.

It was time to do some homework. In the geology library, he learned that rocks of the right age were also not far away in Nova Scotia, and, better yet, they were pounded by some of the highest tides in the world, giving him plenty of exposed areas to search. Jenkins supported the expedition, and Neil brought several more experienced hands to explore the seacoast in the Bay of Fundy. They found many fossils, including a tiny jaw with unusual teeth that had characteristics of those of both reptiles and mammals. It turned out to belong to a creature called a trithelodont, a transitional fossil between reptiles and mammals—a great find. The next year, Neil and his team collected three tons of fossil-bearing rock that contained thousands of teeth and bones of these creatures, crocodiles, and lizard-like reptiles.

Tracking the origin of mammals from reptilian ancestors gave Neil a particular interest in some of the great evolutionary transitions in life's history. A second major milestone, the origin of four-legged vertebrates from fish ancestors, soon moved to the forefront.

When Neil joined the University of Pennsylvania as an assistant professor in the late 1980s, he was determined to uncover some of the missing links among some of our earliest backboned ancestors.

Neil and Ted's excellent adventures

The critical period for vertebrates coming onto land is within the late Devonian, from about 385 to 365 million years ago. Before that time, vertebrates were represented only by fish. By the end of the Devonian (about 359 million years ago), however, life on land was changing in dramatic ways: vertebrates had evolved limbs and started to walk, and insects and spiders had also entered terrestrial habitats.

In the vertebrate fossil record, a few key fossils mark some of the stages of the fin-to-limb transition and the origin of four-legged vertebrates (tetrapods). For example, fish such as *Eusthenopteron* and *Panderichthys* are known from about 385 to 380 million years ago and show in their fins some of the same basic bone structures, the presence of one upper "arm" bone and two "forearm" bones, as those found in tetrapod limbs (Figure 10.2). But these ancient fish do not have the equivalent of wrists or fingers. By roughly 365 million years ago, the full suite of tetrapod limb characteristics are present in animals such as *Acanthostega* (Figure 10.2), discovered on the east coast of Greenland. One of the most surprising features of this animal are the eight digits on its hands (and at least that many on its feet), which reveal that the early tetrapods had more than the five digits common to all recent forms.

But as great a fossil as *Acanthostega* is, when Neil began his studies, there was quite a gap between its fully formed limbs and the fins of earlier fish. It was that gap that Neil and his new graduate student, Ted Daeschler, aimed to close.

Fortunately for Neil and Ted, the Commonwealth of Pennsylvania sits on a lot of Devonian rock. Some 380 million years ago, the Acadia mountain range was drained by a series of meandering rivers that emptied into the inland Catskill Sea. The resulting Catskill Delta now lies in today's Appalachians, with ancient floodplain deposits extending from southeastern New York through Pennsylvania, Maryland, and West Virginia. So, luckily, Neil and Ted would not have to go far to go prospecting. But, just as in Connecticut years earlier, the rock is largely covered with greenery and urban development, and there is no seacoast. The best Neil and Ted could do was to scout highway road cuts, which at least had the advantages of being easily accessible and made for inexpensive field trips.

For several years in the early 1990s, Neil and Ted embarked on a series of roadside adventures. On State Route 120, they found a newly blasted exposure of late Devonian rock called Red Hill. They first found a few fish scales there; then, while Neil was off in Greenland on another expedition, Ted returned and found the shoulder of a tetrapod—the first late Devonian tetrapod discovered outside Greenland. Comprising about 75 vertical feet of rock formed from about 400,000 years of Devonian deposits, the hill was loaded with fossils. With such easy access, the two men could haul anything back to the lab at Penn for closer inspection.

Route 15 provided another bonanza. Combing through some freshly blasted boulders, Neil and Ted carted a few big hunks back for further analysis. They spotted a large fish fin poking out of one boulder, but it was not the sort of fin they usually found. This one had bones inside. A month's worth of preparation revealed a fin with one bone attaching to the shoulder, two bones attached to it, and eight rods extending out from the fin. The eight rods looked as if they could be forerunners of digits, the eight digits that appeared in animals like *Acanthostega*.

Their "fish with fingers" and three different tetrapods gathered from road cuts were valuable new fossils, a great bounty for a few years' effort, but they weren't filling that fish–tetrapod gap. Since tetrapods and various fish coexisted at Red Hill, it became clear that the rocks were too recent (Red Hill was 361–362 million years old). The essential transition had taken place some time earlier.

If they were to find transitional fossils, they had to look at rocks that were a bit older. They had learned about the right kinds of rocks from Red Hill and other road cuts. Fossils were best preserved in deposits at the margins or overbanks of ancient streams that were part of delta systems. But where in the world might they find such rocks that others had not already explored?

They considered China, South America, and Alaska, but the prospects were not encouraging. Then one day, in the course of settling an entirely unrelated geological squabble, they opened an old geology textbook and happened on a map showing several late Devonian deposits in North America. It showed East Greenland, but Neil and

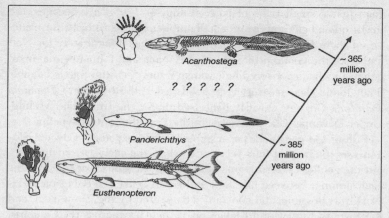

Figure 10.2 **The origin of tetrapods and limbs.** A few of the better known fossils are shown that connect the evolution of fish to the origin of tetrapods. There is a substantial gap in body and limb form, and time, between animals such as *Panderichthys* and *Acanthostega*. Drawing by Leanne Olds, redrawn from P. E. Ahlberg and J. A. Clark (2006), *Nature* 440: 747–49, and N. A. Shubin, E. B. Daeschler, and F. A. Jenkins, Jr. (2006), *Nature* 440: 764–71.

many others had already been there. It showed the Catskill Formation, where they had now toiled for years. And it showed the Canadian Arctic Islands—a vast and, for paleontologists, virgin territory.

Excited, they went off to discuss a possible expedition over lunch at their favorite Chinese restaurant. At the end of the meal, Neil opened his fortune cookie and read: "You will soon be at the top of the world."

A treasure map

If Neil was to find his fortune at the top of the world, he and Ted needed to do their homework. The first order of business was to find the right map. The Arctic Islands encompass over 75,000 square miles and include a number of uninhabited islands in one of the most remote parts of the planet. That is a lot of territory to cover and, because of the extreme climate, they would have a very short time to scout and decide where to dig. They needed to narrow their search.

Fortunately, many of the remote places in the world attract more than just a small tribe of paleontologists. For many decades, the major oil and gas companies and a number of governments have surveyed them for prospective natural resources. Fortunately for Neil and Ted, the Geological Survey of Canada and a host of major oil companies had sponsored an extensive survey of the Arctic Islands. They found their treasure map published in the *Bulletin of Canadian Petroleum Geology*, cleverly disguised under the title "The Middle-Upper Devonian Clastic Wedge of the Franklinian Geosyncline."

Filling 154 pages, the paper by Ashton Embry and J. Edward Klovan was the fruit of four years of work in the early 1970s, mapping out the geological formations of the various islands. For two months each summer between late June and late August, the only time field-work was possible, and constantly hampered by fog, snow, rain, and high winds, Embry and Klovan made their way across the Canadian Arctic by helicopter and light airplane, taking measurements and samples at every step.

Neil and Ted followed Embry's tracks by combing their way through page after page of the descriptions of the geology of the various rock formations exposed on the islands, looking for hints of where to go. Then, buried deep in the paper in the discussion of the so-called Fram Formation, which ran across southern Ellesmere Island, they hit the sentences that had them ready to pack their bags:

> the . . . fossil content of the Fram Fn [Formation] suggest a mean-dering-stream environment of deposition. Sandstone units are in-terpreted to be point-bar and channel-fill deposits, whereas the shale siltstone units *are of overbank origin*.
>
> *The Fram Fn is similar to the Catskill Fn of Pennsylvania* [italics added]. . . .

Embry and Klovan thoughtfully included a representative photo of these deposits (see Figure 10.6, top).

But this terrain was even better than the Catskills, for in this part of the world there was virtually no vegetation covering the rocks, giving them endless exposures to survey.

So the Fram Formation was their target; how would they get there? They figured the best way to find out was to track down Ashton Embry, some twenty-three years after his survey.

Neil and Ted flew to Calgary to meet with Embry and his team of savvy veterans who had been traveling throughout the Arctic every summer. They explained their Catskills work and presented their ideas for exploring the Arctic.

"Great idea. You are going to find what you are looking for," Embry assured them.

The Canadians' expertise in geology and logistics was priceless. Getting around the islands was difficult, for there were very few settlements or airfields. The distances between islands were beyond the normal range of helicopters and required a system of fuel depots so that helicopters could hop from one island to another.

There were also matters of funding and planning the expedition. They soon received a generous commitment from an anonymous donor that would cover all of their costs. Neil's former advisor, Farish Jenkins, had organized many field studies in Greenland, so Neil invited him aboard as a partner. The expedition would then cover three academic generations: Farish, Neil's thesis advisor; Neil, Ted's thesis advisor; and Ted.

By the spring of 1999, plans were well under way for a six-week trip that summer. With the certain uncertainty of the weather, every contingency had to be considered as they bought supplies and worked out the travel logistics. At $2,000 per hour for helicopter time and with limited cargo capacity, they had to be very disciplined about the supplies they brought.

There was also the matter of permits. The expedition was headed into Nunavut Territory on Ellesmere Island, which was controlled by the Inuit people, and permission was required from the local ministry and the hamlet of Grise Fiord. This settlement of 140 people, the northernmost village in North America, would be the team's last destination before venturing into completely uninhabited terrain. Everything was going along as planned until a second group, an association of hunters and trappers, refused to grant a permit, fearing that the team's aircraft would disturb the wildlife. It was a big setback, but they shifted their plans away from Ellesmere Island to Melville Island in the west (Figure 10.3).

Melville Island

The staging point for reaching Melville is a small Inuit hamlet on Cornwallis Island called Resolute Bay. With about 200 inhabitants,

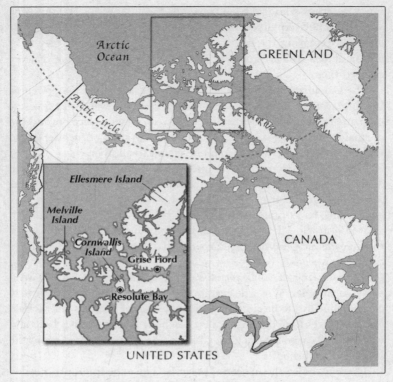

Figure 10.3 The Canadian Arctic Islands. The locations of Elles-mere, Melville, and Cornwallis Islands, and the hamlets of Resolute Bay and Grise Fiord. Map by Leanne Olds.

it serves as an aviation hub and has a grocery store and three hotels —which, as it turns out, was fortunate for the six-person field team, for they were grounded there by bad weather for several days.

They were also a bit unsettled by small talk with the residents. When asked where they were headed, "Melville Island!" was the team's enthusiastic reply. The natives shot back a look of "Oh, you are not really going *there*, are you?" They could not think of any good reason to go to such a desolate place. Neil was not encouraged; "It was like saying we were going to Castle Dracula for dinner."

Time was precious, and they had to go at the slightest break in the weather. When the veteran bush pilot finally got them aloft in his twin-engine Otter, they wished they had waited longer. It was a scary ride, for they were enshrouded in fog the whole way as they looked for a fog-covered island. The pilot made several passes before landing the plane on the tundra. They unloaded their gear, then the pilot wished them luck and was gone.

The first thought on Neil's mind? Survival. And the first order of business? To load the rifles! This was polar bear country.

Then they set up the tents and made camp. They would have twenty-four-hour daylight, freezing cold temperatures, and high winds throughout their stay. They had to brace the tents with rocks. They also set up a tripwire system around the perimeter. Should a polar bear stray into camp, it would trigger an alarm and give everyone a few moments to grab their weapons.

After a long day, they finally settled in for the "night."

Less than an hour later, the alarm went off. Everyone scurried out of their sleeping bags and appeared, rifles ready. It was a false alarm; the wind had blown over one of the poles holding the tripwires. Still on edge, everyone tried to get back to sleep.

Thirty minutes later, the alarm sounded again. It was yet another false alarm. This time a wire had come loose.

Twice more the first night the alarm went off. The team leaders, disgusted, said to hell with it, and turned the whole system off for good.

Beyond worrying about becoming a polar bear's breakfast, Neil had other concerns. Here they were, isolated in a scary, desolate, and very unfamiliar place, for quite a long time. Would they find anything?

The first thing paleontologists want to do in a new location is scout. They hiked to a hill behind the camp and promptly found some fossil fish scales. Good enough, they had reason to hope.

Their only contact with the outside world was by radio. They set up the antenna so they could make their two scheduled calls each day to check in and let their logistics support team know how they were and when the team's help was needed to move to a second camp. They continued to find fossils at the second camp, but it became clear they were looking at deposits of a deep-water marine environment, not the shallow streambeds and overbanks they wanted.

In the fourth week, bad weather set in. For thirteen straight days the wind blew at 30–50 miles per hour. The men were stuck in their tents. Soon they had read all of their own books and those everyone else had brought along for just such a contingency, from Bill Bryson and Carl Hiassen to Tolstoy. Neil then passed the time perfecting pocket rockets—matchhead-powered aluminum foil missiles that, once perfected (and he had a lot of time to perfect them), could be shot twenty feet across the tent.

Then it was time to go home, and they realized that the long expedition had been a bust. If they were going to return to the Arctic, they had to get to Ellesmere Island and the heart of the Fram Formation.

Ellesmere

The team got all the necessary permits in time for the summer 2000 field season. They flew to Grise Fiord, at the southern end of Ellesmere, and went by helicopter to a section of the Fram Formation described by Ashton Embry.

The full field crew this year was a team of nine, including Neil, Ted, and Farish. It took a lot of supplies to house, feed, and equip a crew so far from even an outpost such as Grise Fiord. Everything had to be carried in and out by helicopter. At the same time, the leaders could not be too spartan. The twenty-four-hour daylight, the wind and cold, all the walking, and the hard work would drain the energy of the most enthusiastic fossil prospector. Neil dropped twenty pounds in the course of the season. A few comforts and rituals were necessary to maintain both team spirit and physical strength.

For Neil, a nice martini was his reward at the end of the day. So he packed a martini shaker, plastic glasses, and vermouth. But he was pained to find out that at camp, above the Arctic Circle and just 10 degrees latitude below the North Pole, global warming had destroyed his perfect cocktail—there was no ice! Somehow, he made do.

With considerable field experience among them, the team leaders knew that dinner was crucial to maintaining morale and camaraderie. They packed military MREs (meals, ready-to-eat) for emergencies, but they spent part of the spring dehydrating food in the lab and preparing some tasty meals, complete with sauces and spices, to perk up "Café Ellesmere." The menu, carefully designed to provide a new set of flavors each day, included pasta in red sauce, risotto, white chili with turkey, shepherd's pie, "Parry Islands" gumbo with crabmeat, Tuscan stew, and "Aloo Gobi" curry.

Using just a propane stove and fresh water from a nearby stream, the crew rotated cooking duty. The evening ritual of meal preparation and dinner provided a chance to talk about the day and to plan for the next. After cleaning up and making a radio call to base, card games passed the time until about nine-thirty, when everyone retired to their sleeping bags—in their individual tents (given six weeks under these conditions, they could not risk roommate problems).

Sometimes, however, rituals and even dinner had to be put aside. One day late in the season, one of the undergraduates on the crew, Jason Downs, had not returned to camp by dinnertime. With polar bears in the area, potential bad weather, and the possibility of getting injured or lost, the team was prepared to mount a search. Before they set off, Jason finally appeared, pulling handfuls of fossils from his pants, jacket, and backpack (Figure 10.4). He had stumbled on a trove of fossils only a mile from camp and carried back all he could.

With round-the-clock daylight, there was no need, nor any patience, to wait until the next day. Forgoing dinner, the crew grabbed

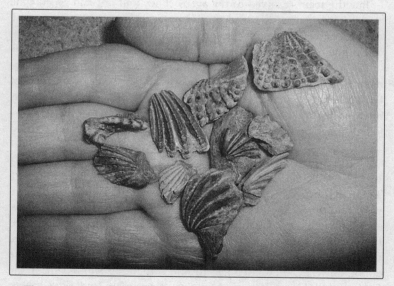

Figure 10.4 A promising handful of Devonian fish fossils. These lungfish tooth plates were scooped up off the surface of Ellesmere Island, near the site where the team decided to dig. Photo courtesy of Ted Daeschler, Academy of Natural Sciences, Philadelphia.

some candy and energy bars and headed off to Jason's site. They crawled all over it, picking up fossils and trying to figure out where and how to excavate it. Digging revealed layers of fossil fish skeletons, definitely more than the teeth and scales they had been finding elsewhere.

But they weren't in Pennsylvania with a pickup truck. The only way to find out what was really there was to encase boulders of fossil-bearing rock in plaster jackets and airlift them back to Grise Fiord, then fly them back to their labs in Chicago and Pennsylvania for detailed examination. They could only bring back a few large jackets.

When the jackets were opened and the rock carefully removed from the bones, a number of different fish were revealed—lungfish, some other lobe-finned fish, and a few placoderms. Unfortunately, all of the fish had already been found in Latvia. This disappointment highlighted yet another risk of an expedition: even if fossils are found in a new location, the fossils themselves may not be new or informative.

But the team resolved to try again. They had not had much time at the new quarry they had dug in 2000, so the next season they would focus on the quarry and prospect some new sites. Returning to Ellesmere in 2002, they excavated five more jackets.

In the lab that following winter, they uncovered an enigmatic fragment of a snout. They had enough of it to say it was some kind of flat-headed animal, maybe even a tetrapod, but they could not determine what it really was. They would have to return to the Arctic a fourth time and find more of the animal.

Tiktaalik

These expeditions were expensive and had little to show for the five years and three expeditions. It cost $120,000 to return to Ellesmere, and funding it was difficult. The team eventually gained support from the National Geographic Society, the National Science Foundation, the University of Chicago, Harvard, and again from their private donor, but they knew they had to find something significant or the funds would dry up. In early July 2004, three team leaders and three crew members set off for another six weeks of summer in the Arctic.

The first day in camp, the wind was blowing very hard, so Neil decided to eat his lunch on another slope. As he sat down, he noticed that the rock was covered with what appeared to be bird droppings. But it wasn't guano; the white spots were fish scales, lots of them. Neil probed around and found a piece of a jaw of something that looked like *Panderichthys*. That was encouraging.

The team opened the quarry with renewed vigor. They removed a few feet of protective gravel they had left over the site two years earlier, and each person started working at a different level along the slope. Neil was working at the bottom of the quarry in sediment that was still frozen. He found another patch of scales unlike any he had seen before and started to remove the fossil material and surrounding rock. As he kept digging through the ice, he found a set of jaws —also unlike any fish jaws he had seen before—perhaps the kind of jaws, he thought, that might connect to a flat head.

The next day, Steve Gatesy was working at the top of the quarry, just six feet above Neil. As he popped out a piece of rock a snout appeared, looking right at him (Figure 10.5, top). The animal had a flat head. Better yet, since it was facing out, it suggested that there might be more of the skeleton embedded behind more rock. Steve then carefully removed as much of the surrounding rock as possible so the fossil could be jacketed.

The team was being pounded by rain, sleet, and snow, but they could not have cared less. They were pretty sure that they had something new and kept working. Near the end of the season, Farish found another specimen, the largest of the three thus far. All three were jacketed in plaster and flown home.

Only so much about a fossil can be figured out in the field. The team was excited, but the moments of truth would be in the prep labs after the jackets were opened and as the rock was carefully removed from the bones, often using dental picks. The preparators, Fred Mullison in Philadelphia and Robert Masek and Tyler Keillor in Chicago, went to work, gradually revealing what lay in each jacket. Photos were sent between the labs while Neil and Ted spent several hours each day on the phone. Day by day, week by week, more parts of the creature emerged, and after two months of meticulous work, what a creature it was (Figure 10.5, bottom).

It had scales on its back, like a fish. But unlike fish, which have conical heads, it had a flat head, like a crocodile's. And that head—unlike fish heads, which are connected directly to their shoulders—was connected to the trunk by a neck, like those of four-legged animals. It

Figure 10.5 *Tiktaalik* **emerges.** Top, the snout of *Tiktaalik* protrudes from the rock (arrow). Here, team members work to remove the surrounding rock so the specimen can be jacketed and shipped for detailed preparation. Bottom, *Tiktaalik* in full glory. Note the presence of a neck, the eyes on top of the head, and the fins with bones inside. Photos courtesy of Neil Shubin and Ted Daeschler, respectively.

had fins with webbing, but it also had bones inside the fin that correspond to the upper arm and forearm, and, most unexpectedly, it had a wrist bone—something not found in any other fish. Moreover, the fin bones had joints that allowed them, like our limb bones, to move and flex and extend. This fish could do push-ups.

It was part fish, part tetrapod—a fishapod.

It was exactly the sort of intermediate between water- and land-dwelling vertebrates the team had hoped to find, but their quest turned out far better than they could have imagined. They had more specimens, and more complete and exceptionally well preserved specimens, than they could have reasonably wished for, the largest representing an animal nearly nine feet long. And all of them had been found in one location, in rocks of the right age (375 million years old), and in exactly the sort of ancient stream environment they had anticipated.

Ashton Embry was right—they would find what they were looking for. But he was even more right than he knew. When the team took photographs of the quarry slope, they realized that they were digging at the very spot that Embry had photographed thirty years earlier, which they had seen in his paper (Figure 10.6). It really did pay to do one's homework.

As the discoverers, the team was entitled to name the new creature. They decided to give it an Inuit name, both to reflect its origin in the Nunavut Territory and to thank the people who had granted them access to their lands. They consulted with the Council of Elders, who suggested two possible names in the Inuktitut language. The team went with *Tiktaalik* ("large freshwater fish") as the genus name and *roseae* as the species name (after their first benefactor).

A star is born

Tiktaalik roseae made its public debut on the cover of the journal *Nature* on April 6, 2006. The team described the discovery, the animal, and the insights it offered into the origins of tetrapods and limbs in a pair of articles, and the paleontologists Per Ahlberg and Jennifer Clack offered an enthusiastic commentary. They immediately drew a comparison to *Archaeopteryx* and ventured that *Tiktaalik* may well become a similar icon of a major evolutionary transition.

The mainstream press was equally if not more taken with the dis-

Figure 10.6 Déjà Vu in the Fram Formation. Top, photo by Ashton Embry of Fram Formation rocks on Ellesmere Island, taken in July 1974 and published in his survey of Arctic geology. Bottom, photo of quarry location where *Tiktaalik* was found (a team member is standing at the quarry edge), taken in July 2004. They are the same location. Photos courtesy of Ashton Embry and Neil Shubin, respectively.

covery. It was reported "above the fold" on page 1 of the *New York Times*, and feature stories ran in *Time* magazine and on the major TV networks' evening news shows. Arriving in the midst of yet another wave of the long-running creationist battle against evolution, the 375-million-year-old creature that was so obviously transitional between fish and land animals was a most welcome and potent blow to the skeptics' rhetoric about the purported lack of transitional forms in the fossil record.

While the media's tendency to characterize *Tiktaalik* as a "missing link" was understandable, Neil Shubin, preparing to return to the Arctic to find more Devonian treasures, set the record straight:

When people call *Tiktaalik* "the missing link," it implies there is a single fossil that tells us about the transition from water to land. *Tiktaalik* gains meaning when it's compared with other fossils in the series. So it's not "the" missing link. I would probably call it "a" missing link. It's also no longer missing—it's a found link. The missing links are the ones I want to find this summer.

PART THREE

THE NATURAL HISTORY OF HUMANS

The discoveries of Tiktaalik and Archaeopteryx, Deinonychus, and feathered dinosaurs filled gaps in the sequences of fossils, marking transitions from one kind of animal form to another. The bodies and key body parts of these fossils have provided critical clues to the origin of tetrapods from fish and to the origin of birds from theropod dinosaurs. To reconstruct human origins from our ape ancestors, a similar sequence of hominid fossils would be necessary.

Dubois' discovery of Java Man (Homo erectus), as fortunate and timely as it was in providing evidence that intermediates existed between humans and the common ancestor we shared with the great apes, provided just a single point on our evolutionary line—a line whose length we knew nothing about at the time. Moreover, Homo erectus was very human-like in certain features, such as its thigh bone, and its other skeletal features were entirely unknown. Furthermore, Dubois did not find any artifacts (e.g., tools) that indicated what behaviors Java Man was capable of. To understand the sequence of human evolution, links were sought in two directions from Homo erectus—back in time toward the common ancestor of apes and humans and forward to the dawn of modern humans.

But the recovery and analysis of the hominid fossil record have presented many challenges, particularly in the early days of paleoanthropology. These problems are perhaps best understood by contrasting the approaches available to or taken by the scientists in the previous stories. Success in finding fossils depends on many factors. Neil Shubin's team succeeded because they located deposits of the right age (late Devonian) in the right kind of environment (stream overbanks) that were exposed and accessible. Theirs was a well-educated guess, and if they got lucky, theirs was a form of smart luck (although finding several good-sized specimens in the condition they did was remarkably good fortune). But, in terms of finding the proverbial needle in a haystack, they knew what kind of haystack they wanted and, thanks to prior geology, they knew where it was.

Similarly, one might describe Walcott's discovery of the Burgess Shale as lucky, but with forty-five years of experience in geology and in collecting Cambrian trilobites, hitting a mother lode of Cambrian animals was probably more a matter of justice than luck. But keep in mind that marine creatures such as trilobites, crustacea, molluscs, and many other groups existed in huge numbers and were widely distributed. They weren't rare, solitary animals. Furthermore, they were buried in seabeds that preserved many of them well. Their numbers and lifestyle ensured that their fossil record would be abundant.

Now, contrast these finds with Dubois' and Roy Chapman

Andrews' strategies and experiences. Dubois had some sound reasons for going to Sumatra and Java, but one look at his veranda (Figure 5.4) reveals that he had found haystacks full of fossils but had no way of knowing if it was even possible that there would be a few needles of hominid among the tons of other beasts. Imagine trying to distinguish a fragmentary shard of hominid bone in a field of other bones. Human skulls, femurs, and teeth are distinct, and that is what Dubois was able to identify. Roy Chapman Andrews, following Dubois' lead into Asia, found tons of large Cretaceous dinosaur and Eocene mammal bones but not a trace of ancient humans. Those rocks were the wrong age. It is worth keeping in mind why our museums (and their storerooms) are full of dinosaurs—they left behind big bones, which are easy to spot and less easily destroyed, and they thrived for a long geological time span.

Finding hominids was and is an entirely different quest. Our ancestors did not roam in herds across vast areas or live in seabeds. They were not numerous and were much more restricted in their distribution in time and space. Their skeletons were readily separated from their skulls, and the latter typically got smashed to bits. For nearly forty years after Dubois' discovery, including Andrews' expedition, no new ancient hominids significant to the human–ape link were found. Neanderthal fossils *from Europe were even more like modern humans in terms of build and brain size; it was debated whether they were even a distinct species. They certainly did not come close to bridging the gap between humans and the great apes. Other links in the ape–human sequence were not known or recognized. So, in the beginning of the 1920s, our knowledge was literally stuck in pretty much the same place that Dubois had left it. How could one find ancient hominids?*

Various kinds of bias played a role in paleoanthropological pre-conceptions about where ancestors would be found and how old they were. Thanks to Dubois, most thinking about "the Cradle of Mankind" was focused on Asia. Naturally, some followed Dubois to Indonesia, some prowled around China. The continent was also the home of distinct populations (races) and the oldest human civilizations known at the time, so many thought the most imme-diate ancestor of modern Asians would have originated there. Meanwhile, the abundant fossil record of Neanderthals around Europe prompted some to think that modern Europeans had a separate origin from that of Asians, from a Neanderthal ancestor.

Thus, despite Darwin's hunch about Africa and human origins, attention was focused elsewhere. In 1924, the anatomist Raymond Dart reported the discovery of a striking fossil skull in a quarry in South Africa. He named the creature Australopithecus africanus *("southern ape from Africa") and saw it as an intermediate between apes and humans. Because the skull was impossible to date and Dart lacked any evidence of tool use, his claim was rejected by the leading authorities, who saw it as "just an ape." The creature's origin did not fit the bias at the time of what a hominid ancestor would look like or where it existed.*

But the pendulum was going to swing back to Africa. The main catalyst at first was not fossils but tools. Beginning in the late 1920s, many loads of tools were unearthed in East Africa, as old or older than those found elsewhere, that indicated the presence of ancient toolmakers. The herculean effort of finding those tools and the fragmentary remains of their makers is the heart of the first story that follows. It was not for three more decades, until 1959—exactly 100 years after The Origin*—that new hominid fossils were found in East Africa and that the picture of the sequence of hominids that link us to the great apes started to be revealed.*

The focus of human origins has remained in Africa ever since, but the emerging picture of human natural history has been drawn from more than fossils and tools. Entirely new and unforeseen ways of deciphering human history from the DNA of living and ancient humans have revolutionized the science of, and rewritten the story of, human origins.

Figure 11.1 **Olduvai Gorge.** This gorge contains a rich record of the last 2 million years of human history. Photo from M. Leakey (1984), *Disclosing the Past* (Doubleday and Company, Garden City, N.Y.). Leakey Family Archives.

⁓ 11 ⁓
Journey to the Stone Age

> Man is a tool-making animal.
> —*Benjamin Franklin*

HIS KIKUYU NAME was Wakuruigi—Son of the Sparrow Hawk.

His parents called him Louis.

Although he was the son of English missionaries to Kenya, Louis Leakey considered himself more Kikuyu than British. He was born two months prematurely, in Kenya in 1903, and his first great achievement was to survive. At the time, and especially in rural Kenya, no facilities were available for vulnerable infants. His "incubator" was a shroud of cotton and wool and a bed heated by a charcoal fire.

Yet survive he did, aided no doubt by the customary spit of the natives to ward off "the evil eye." Although he was the first white baby that most had ever seen, Louis was welcomed into Kikuyu society. Kikuyu boys are organized into groups by age, and at eleven, Louis was accepted as a member of the Mukanda group ("the time of the new robes") and prepared for the secret rites that would mark his passage into manhood.

Like his Kikuyu "blood brothers," Louis built himself a "bachelor's quarters"—a mud hut where he could live apart from his family. By the age of fourteen he built, slept in, and lived in his own three-room house. As he later explained, "I did everything except feed; for my parents insisted that I should still feed with the family, which was essentially sound."

Louis learned the ways of the Kikuyu at first hand. His Mukanda brothers taught him how to throw a spear and wield a war club. An elder, Joshua Muhia, taught him how to track, hunt, and trap wild game. Louis learned the habits and signs of the dik-dik, duiker, mongoose, jackal, hyena, and virtually every other animal in the vicinity. Most important, Joshua also taught him to be a patient observer.

Louis's hut was soon filled with a collection of bird eggs and nests, animal skins, bones, and interesting stones. His older sister thought this personal museum was dreadful, but Louis could not have been happier.

A prominent and beloved missionary, Louis' father, Harry, often received English guests. To the boy's good fortune, they included Arthur Loveridge, the first curator of Nairobi's Natural History Museum. Loveridge liked Louis and encouraged his collecting and his interests in natural history. He taught Louis how to prepare specimens and about classification. Louis, in turn, provided the museum with three snakes, a live horseshoe bat, a porcupine, a melanistic form of a genet cat, and a variety of birds.

Louis loved life on the mission, Kenya, and natural history. He was also quite devout. He thought he would grow up to be a missionary himself, with ornithology as his main hobby.

But few experiences stoked Louis' interests, or had a greater impact on his future, than a simple children's book he received as a Christmas present from a relative in England. Entitled *Days Before History*, it recounted the tale of a young boy named Tig in the Stone Age of Britain. Descriptions and drawings of the flint arrowheads and axes of this culture captivated Louis. Convinced that Stone Age people must have also lived in his region of Kenya, he began collecting pieces of rock with any resemblance to the flint tools he read about. His family teased him about what they called his "broken bottles."

One day, Louis timidly showed his collection to Arthur Loveridge, who confirmed that some of his stones were indeed ancient tools. Louis was overjoyed as Loveridge explained that Kenyan Stone Age cultures made tools out of obsidian, as flint was not available, and that he could show him some examples from the museum. Louis became obsessed with collecting stone tools. As Loveridge had taught him, he kept good records of each find and pored over the few books available on the subject. Louis learned how little was actually known about the Stone Age, particularly in East Africa, and he made up his mind that this would be his quest. He was thirteen years old.

A good kick in the head

Pursuing the history of early humans would require, Louis recognized, more formal schooling. He had received relatively little while growing up, just brief spells in British schools during the customary home leaves English missionaries took every few years. Louis was not fond of school and bridled at the restrictions it placed on his physical freedom and limitless curiosity. Fortunately, his parents' leave was delayed by World War I, so Louis spent all of his early adolescence romping about Kenya, exploring caves, collecting tools, and learning Kikuyu ways.

Louis hoped to attend Cambridge University, but when he enrolled at Weymouth to start his preparations, he discovered just how far behind his classmates he was and how far out of place this white Kikuyu was in England. Louis worked hard to catch up and to learn how the British system worked.

He discovered several obstacles between himself and a Cambridge degree. First, there was the matter of money. He and the Leakeys had very little, a condition that persisted throughout his life. Second, if Louis wanted any kind of financial assistance, he would have to pass several subjects in an entrance exam. And third, he needed to demonstrate a mastery of various subjects to meet the requirements of the bachelor's degree (the Tripos).

As Louis read the *Student's Handbook*, he thought of a solution to all of these obstacles: he would offer Kikuyu as one of the modern languages for his scholarship exam and for part of his Tripos. The university agreed that Kikuyu was a modern language, but no one on the faculty knew it! Louis proposed that he would submit a "certificate of competent knowledge" from a reliable source; who could be better than Chief Koinange of the Kikuyu tribe? The chief signed the document with his thumbprint.

Louis was getting the hang of academia.

He hurled himself into university life, perhaps with a bit too much gusto. Despite his inexperience in conventional athletics, he was determined to make the rugby team and to represent the university in its annual match against rival Oxford. During a trial match early in his second year, Louis was trying to impress the team captain when he was kicked in the head and had to be carried off the field. As soon as he regained some of his senses, he reentered the game and was hit again and put out for good. He had a crushing headache that got worse over subsequent days. He was so debilitated that he could not do his schoolwork.

The doctor told him to rest for ten days, but Louis discovered he had lost some of his memory, and the headaches returned as soon as he tried to work. He turned out to have a form of post-traumatic epilepsy. His doctor insisted that he take a year off, preferably somewhere in the open air.

This seemed to be a terrible setback to Louis' ambitions to get a degree quickly and to launch his archaeological career in Kenya. But, as luck would have it—the kind of luck that Louis would enjoy again and again in the coming years—he did find a worthwhile task to pursue during his recovery.

Through a family friend, Louis heard that the Natural History Museum in London was planning to send an expedition to the Tanganyika Territory (modern Tanzania) to collect dinosaur fossils and they needed someone with experience in Africa. (The leader of the expedition, a man named Cutler, had never been to Africa.) Though just twenty, the young man got the job and sailed, first class, for East Africa.

Louis quickly proved invaluable, for he was adept at organizing an expedition. After helping Cutler navigate the Tanganyika customs office and determine when his equipment would arrive from England, he set out ahead to locate the field site, called Tendaguru, and to build a camp there. In a coastal port town, a fortunate coincidence led to his meeting the headman of a village in the area of the Tendaguru district; he agreed to help Louis find the site and to recruit native men to build the camp. Louis secured fifteen porters, a cook, a gun-bearer, and a young Kikuyu assistant, and he and the village headman started out on foot on what he thought was "a great adventure."

They covered over fifty miles in three days before reaching Tendaguru Hill. The headman asked Louis to fire his rifle in the air so the natives would know that a white man was in the district. Many villagers came to offer various staples, which Louis purchased. The headman then beat out a long message on a signaling drum, explaining that he was telling villagers within earshot to come the next morning to greet Louis and to bring their knives and axes to help build a house. That night, Louis settled down with "a strange mixed feeling of pleasure, triumph, expectation, and loneliness."

The house was built in a few days, and Louis went on to build storehouses and a kitchen and to find a supply of water. He made several more treks to and from the camp and the coast before Cutler joined him two months later. They spent four months together in the field.

It was not exactly what the doctor had in mind when he ordered Louis to rest, for it was hard work, but Louis loved it, and he learned a lot. At five o'clock every morning except Sunday, he marshaled the workmen who helped with the excavation. He accompanied Cutler to dig sites and spent the day assisting him in encasing the bones in plaster of paris. Louis found Cutler to be an excellent teacher, and he gained great practical training in fossil excavation and preservation. The conditions were very difficult. It was very hot, water was scarce and had to be fetched from many miles away, and Louis had close encounters with various animals—buffalo, elephants, a mamba, and a leopard—that could have been the end of him. Disease was also widespread; Louis went down with dysentery and malaria at the same time.

Regardless, Louis wanted to stay longer, but he had to return to Cambridge. He later wrote: "I little thought when I was kicked in the head what a great effect that incident was going to have on my whole career."

Once Louis left, Cutler struggled. He could not manage the workmen very well and he was plagued by tropical ailments. Nine months later, he died in camp from complications due to malaria.

Back to the Stone Age

With his Tendaguru experience, Louis' confidence grew enormously. Always looking for ways to make a few extra pounds, he started giving public lectures, on "Digging for Dinosaurs." His talent for exciting audiences would later be displayed on the world stage.

Louis was eager to resume his pursuit of the Stone Age in Kenya. He split his remaining time in Cambridge between his formal studies, examining prehistoric sites in Britain, learning the art of making stone tools, and investigating the history of African bow and arrow technology. Taking his final exams, he passed with high honors.

It was time to put his learning into practice. Firmly convinced of the theory of evolution and the antiquity of man, he was certain that the true early history of humans differed from the biblical account.

However, what little was known of that history did not seem to point to Africa. Darwin, in *The Descent of Man* (1871), had stated: "It is somewhat more probable that our early progenitors lived on the African continent than elsewhere. But it is useless to speculate on this subject." By 1926, however, very few scientists thought man had originated in Africa. Dubois had found *Pithecanthropus*, the earliest hominid fossil known at the time, in Asia. Roy Chapman Andrews' expeditions were looking for more hominids in Mongolia. And soon Davidson Black and Wenzhong Pei would find more ancient hominids in China, which would reinforce the "out of Asia" theory. One professor had told Louis not to waste his time searching for early man in Africa, since "everyone knew he had started in Asia."

Nevertheless, Louis knew that the tools he had found as a boy meant there had to have been toolmakers, early humans, in Africa. Setting out for Kenya as soon as he could, he recruited another Cambridge student to work with him, and the two constituted what Louis dubbed the first "East African Archaeological Expedition." This "team" was expanded in Kenya by several of his Mukanda brothers and his actual brother Douglas. Their first excavations were in places Louis knew from childhood. Since caves were the source of many artifacts in Europe, Louis focused on the many caves and cliffs around the Rift Valley near his home and on the lands of settlers he knew.

He hoped to find tools and, perhaps, the bones of those who made them. At the time, archaeologists believed that the oldest culture was the so-called Chellean, named for hand axes from Chelles in France. He figured that if he could find tools of Chellean type or age in East Africa, he would have his evidence of early man in Africa.

His team excavated several caves, rock shelters, and cliff faces in the area near Lake Elementeita. He found, close to or protruding from the surface, tools, shards of pottery, and human skeletons that had been buried deliberately in some sort of funeral ritual. Louis determined that these were the remnants of a prehistoric "Elementeitan" culture. In a year's time, he collected over a hundred crates of specimens, even though his was more of a reconnaissance expedition, to find places worthy of further exploration. Altogether, he found nearly seventy promising sites. He also met another promising prospect, a young woman named Frida Avern. Despite considerable differences in background—she was the conventional daughter of a prosperous British merchant, he, an untamed white Kikuyan—each was taken with the other. Within a year they were married, and Frida then accompanied Louis into the field for his Second East Africa Archaeological Expedition.

His second team was expanded to include a geologist and several undergraduates, whom Louis trained in the field. Their thorough explorations of the richest sites often yielded hundreds of perfectly preserved tools *per day*. Under Louis' tireless guidance, the team worked from dawn to dusk, and Louis continued well into the night, writing up his notes on the day's finds. At one cave, the deepest deposits were rich with artifacts of its prehistoric occupants. Among the bounty of tools were awls made from bird bones, ostrich eggshell beads, and two shards of pottery. The latter would not seem so spectacular amid all of the more numerous and beautiful implements, but this pottery was older than anything that had been found elsewhere at the time.

How old was impossible to say. Accurate dating techniques did not yet exist, and the geology necessary to place these or other sites in a relative time frame had not yet been undertaken. Based on the resemblance of his Elementeitan tools to some found in Europe, Louis guessed this culture was from around 20,000 B.C. (He was often accused of wanting artifacts to be older than they were; in fact, later analysis would date these materials to around 6000 B.C.)

The presence of artifacts at the deepest level of the cave meant it was well worth continuing to look for older sites and artifacts. Late in the season, the geologist John Solomon took a walk along a river gully at Kariandusi and picked up a piece of green lava that looked as if it might be a tool. Louis had no doubts it was a hand ax and sent Solomon and a student back to find all they could. They found plenty. The hand axes were very similar to the oldest that had been found in Europe (Figure 11.2). This was the evidence of early man in Africa that Louis had been looking for.

How early was again difficult to say. For decades, geologists had relied on extrapolations of sedimentation rates to establish a time scale for life on Earth. The idea was that sediments accumulated at a steady rate, so if one measured the depth of the sediment around some ob-

Figure 11.2 Hand ax. Found by Louis Leakey's team at Kariandusi, 1929. Later dated around 500,000 years old, this was the first sign of humanity's long history in East Africa. Photo from L.S.B. Leakey (1931), *The Stone Age Culture of Kenya Colony* (Frank Cass and Company, London).

ject, one could estimate how long ago it was buried. The trouble was, the method did not account for erosion or variation in sedimentation rates. From these estimates, it was thought at the time that the dinosaurs had died out just 10 million years ago (not 65 million years, as we now know), Earth was on the order of a few hundred million years old (versus 4.5 billion years), the most recent ice age (Pleistocene) spanned some 600,000 years (instead of 1.8 million years), and human evolution spanned some fraction of the Pleistocene.

Based on this sort of relative calibration, Louis estimated the tools were 40,000 to 50,000 years old, which was again very old in the context of what others thought at the time. Little did he know that they were actually closer to 500,000 years old. Nevertheless, he was pushing the Stone Age in Africa further back in time and the archaeological and anthropological world was beginning to take notice. After Louis presented his discoveries at a conference in South Africa, sixty scientists followed him back to see his sites in Kenya. They were impressed, convinced that the Stone Age history in Africa did go further back than once thought.

Just how far back did that history extend? Did humans originate in Africa? If so, when? And when did our ancestors first walk upright or first use tools? These questions would require most of the rest of Louis' life to even begin to answer, and many of the most important clues would come from one gorge.

Olduvai

Louis returned to England to study his specimens. He had enough material for two books, one on the tools and other artifacts of the Stone Age culture, one on the people whose skeletons he had uncovered. While he was writing, he planned yet a third expedition.

This time he was determined to scout a place where he thought he might have a better chance of finding fossils. The Masai called it *ol duvai*, "the place of the wild sisal." A long gorge in the heart of the Rift Valley, it was near Ngorongoro crater in the Serengeti Plain of Tanganyika Territory. Louis' attention was drawn to Olduvai by the reports of a German geologist, Hans Reck, who explored it in 1913 and found no tools but many fossils, including those of a modern-looking human. Reck was eager to return, but World War I intervened, and Tanganyika came under British authority. Louis, however, befriended Reck on a visit to Germany and jovially bet him that, should he ever make it to Olduvai, he would find Stone Age implements within twenty-four hours. Now, six years later, Louis invited him on the third expedition.

Louis, Reck, and a support crew of eighteen set out by car and truck on the long trek to Olduvai in late September 1931. They would cross some very rough terrain to reach the gorge, which lay 200 miles from the nearest post office, garage, or trade store. Water was scarce, and the vehicles overheated and consumed much of their precious supply. They found a pool which "contained a liquid con-

sisting of rain-water diluted with baboon's urine . . . quite undrink-
able, but it was usable in the radiators."

For Reck, it was an emotional experience to return to a land he
thought he would never see again. Louis gave him the honor of being
the first of the team to step foot in the gorge. The first night, they
camped under the watchful eyes of hyenas and lions.

Louis was too excited to sleep, so just after dawn's first light he
set out to scout the gorge. He soon found a perfect hand ax lying in
the sediments. Louis, "mad with delight," rushed back with it into
camp and rudely awakened his sleeping team. One of his major goals
of the expedition—to find any trace of prehistoric culture in the rich
fossil deposits of the gorge—had already been met in the first day.
Indeed, the gorge was strewn with such evidence. They found sev-
enty-seven axes in just the first four days. Moreover, some tools were
found in association with fossils of extinct mammals, such as the
great elephant-like *Deinotherium*. They also found 470 hand axes
mixed in with a nearly complete skeleton of an extinct hippo.

The abundance of tools in the five main geologic levels ("beds") of
the gorge was remarkable, for it gave Louis, and soon the world, an
unprecedented picture of an evolutionary sequence of tool technology, including tools older than any found before.

Despite encounters with lions, being charged by a rhino, and having to shoot at a leopard that stalked him, Louis rightly declared
Olduvai "a scientist's paradise."

He won more than the bet with Reck; he found a treasure trove
that would last a lifetime.

Something about Mary

Louis returned again to England, virtually penniless. Living from
grant to grant, he never had any money left over. To make a little
cash, he took advantage of the excitement over his discoveries to
write a popular book about the Stone Age, *Adam's Ancestors*. At the
same time, he had a fellowship to study the artifacts and fossils he
had brought back from Africa. He worked like a fiend and spent very
little time at home. The differences between Frida and himself were
showing, and their unhappy marriage was unraveling.

When Louis needed someone to sketch his specimens, a colleague
introduced him to Mary Nicol, a young artist who had just illustrated

a book. Louis thought her drawings of stone tools were the best he had ever seen and he recruited her for his book.

Mary had archaeology in her blood. Her great-great-grandfather was John Frere, who in 1790 found a number of stone implements near Hoxne, in Suffolk, England. Frere identified these as "weapons of war, fabricated and used by a people who had not the use of metals." Finding them beneath 12 feet of soil, gravel, and a layer of shells led him to conclude that they belonged to "a very remote period indeed; even beyond that of the present world." When Frere made his report to the Society of Antiquarians, there was little sense of how remote that period was. Though overlooked for decades, Frere was later lauded as the first to recognize the relics of a Stone Age culture.

Mary's father was a landscape painter, and the family lived a fairly nomadic existence traveling through Switzerland, France, and Italy. They made several longer stays in small villages in the South of France, where her father befriended an archaeologist who was excavating caves that contained various paintings and engravings. The scientist encouraged Mary and her father to scavenge through his piles of debris, where they found and pocketed all sorts of flint tools. Mary caught the same fever for the Stone Age and its artifacts that had infected Louis as a youngster.

Mary, like Louis, also struggled with formal schooling. She was expelled from one school for refusing to read poetry to an assembly and from another for faking fits in a classroom (using soap to produce a froth) and for setting off a loud explosion in a chemistry class. She later wrote in her autobiography, "At least I ended my school career with a bit of a bang."

Despite never having passed a single school exam, Mary managed to learn about archaeology and assist several digs. Her unconventional upbringing, rebellious streak, independence, artistic talent, and passion for prehistory were irresistible to Louis. And Louis' knowledge, energy, excitement, and attention drew Mary to him. The two fell in love, and Mary accompanied Louis on his next expedition, where she fell in love with Africa as well.

Her first journey to Olduvai etched indelible images of what was to become her home:

When we finally reached the top . . . we could look down into the caldera itself, the Ngorongoro crater, two thousand feet below. The great circular area is some twelve miles across and always densely populated with game, while the shallow soda lake that occupies one small part of it is often pink-fringed with flamingo. If any game can be made out with the naked eye from above, it is elephant, rhino, or possibly buffalo if they are near, but with field glasses one can see animals by the thousand.

. . . starting the descent from Ngorongoro to the Serengeti . . . I was looking spellbound for the first time at a view that has since come to mean more to me than any other in the world. As one comes over the shoulder of the volcanic highlands to start the descent, so suddenly one sees the Serengeti, the plains stretching away to the horizon like the sea, a green vastness in the rains, golden at other times of the year, fading to blue and gray. Away to the right are the Precambrian outcrops and an almost moon-like landscape. To the left, the great slopes of the extinct volcano Lemagrut dominate the scene, and in the foreground is a broken, rugged country of volcanic rocks and flat-topped acacias, falling steeply to the plains. Out on the plains can be seen small hills . . . the scale is so vast that one cannot tell that the biggest is several hundred feet high. Olduvai Gorge can also be seen. Two narrow converging dark lines, softened by distance and heat haze, pick out the Main Gorge and Side Gorge . . . I shall never tire of that view, whether in the rains of the dry season, in the heat of the day or in the evening when one is driving down straight toward the sunset. It is always the same; and always different.

Together they crawled through the gorge, scouring every inch for tools, a hint of a toolmaker, or other fossils. Every productive site was dubbed a *korongo*, which means "gully" in Swahili. It was at Korongo that Mary found two fragments of a hominid skull, with hand axes nearby. It was a tantalizing find, but an excavation turned up no more pieces of the hominid. Nevertheless, Louis was buoyed, writing in his monthly field report that "I still am convinced—that somewhere at Olduvai we shall sooner or later find the fossilized remains of the men who made the Chellean and Acheulean tools which are so plentiful."

Every exposure yielded some sort of archaeological or geological excitement, including a delicate pig skull, a herd of gazelle-like skeletons, and the massive bones of an elephant. Tipped off by a Masai that more "bones like stone" could be found nearby at a place called

Laetolil (now Laetoli), reconnaissance confirmed abundant fossils there. There were also quite live inhabitants as well. One morning, Mary nearly stepped on top of a sleeping lioness. She later explained: "She and I were mutually horrified and fled in opposite directions . . . meeting a lioness on foot is not as disastrous as it may sound, unless she has her cubs with her."

Louis also took Mary to Kondoa to see the paintings on various rock shelters. Mary was entranced by the beautiful human and animal figures and made many tracings of them. The rock art, the dramatic landscape, the beautiful wildlife, the fascinating people—the whole continent had "cast a spell on her."

After an exhilarating nine months, Louis and Mary returned to England, where they married and plotted their next adventure in Africa.

A Stone Age factory

Returning to Kenya, the Leakeys were nearly broke. In order to continue their fieldwork and life in East Africa, they needed some means of support. Louis, already a prolific writer, committed to writing a history of the Kikuyu tribe. Of course, he was determined to make it "the most complete record of a tribe that has ever been written," and he soon figured that volume one alone would run a thousand pages.

Mary just wanted to excavate. Anything would do; she was not concerned about the age or place, and she hurled herself into several sites in the Nakuru region.

The partners' roles shifted. Louis wrote more, lectured, and took up administrative duties for a Kenyan museum, all of which helped to pay their expenses and to create a space for their collections. Mary loved the field and relished every day spent excavating, and living in a grass hut in the bush satisfied her sense of adventure. For security, she obtained a dalmatian, and she so loved the breed that from that day forward she would never be without one or more dogs. She also developed a love for whiskey and Cuban cigars as small rewards at the end of a long day.

Mary quickly proved herself an extraordinary archaeologist, much more methodical and meticulous than Louis. From one trench alone she recovered and catalogued more than 75,000 tools from the late Stone Age.

These first projects wound up with the arrival of World War II. Louis became a civilian intelligence officer and roamed over Kenya, keeping a watchful eye on political developments and scouting for prospective dig sites. Then, children arrived. Their first, Jonathan, was born in 1940. Louis and Mary could then manage only short forays into the field.

On Easter Weekend 1942, they journeyed to Olorgesailie, about 30 miles southwest of Nairobi. Two decades earlier some tools had been reported in the area, but the details and location were sketchy. Louis and Mary, accompanied by a few assistants, spread out over the white sediment. At almost the exact same moment, they called out to each other. Mary kept shouting at Louis to hurry over to see what she had found. He marked his place and went to her. "When I saw her site I could scarcely believe my eyes. In an area of fifty by sixty feet there were literally hundreds upon hundreds of perfect, very large hand axes and cleavers" (Figure 11.3). Mary thought the place looked as though it had only just been abandoned by the toolmakers. Louis

Figure 11.3 Tool factory at Olorgesailie. Mary and Louis Leakey discovered a field of hand axes and choppers at Olorgesailie. Photo courtesy of the Human Origins Program, Smithsonian Institution.

guessed the "factory" site was 125,000 years old, a figure that was scoffed at as too great. Subsequent radiometric dating put its age at more than 700,000 years old.

The scene was so startling and impressive that they decided to leave a large section exactly as they had found it. A catwalk was built, the site was opened as a museum in 1947, and Olorgesailie remains a national museum today.

There was no time to explore further that first day, for they were due at the babysitter's at five o'clock. It was not until the next year that a proper excavation began. Louis was too tied up, so Mary camped at Olorgesailie for several months, with her toddler in tow. It was pretty rough country, water was scarce, and dangerous game was prevalent, but it was now routine for the Leakeys. Mary pioneered the excavation of occupation sites of ancient humans. Archaeologists once thought that such sites would be destroyed by natural processes over time, but Mary's exhaustive excavations of individual layers revealed tools, animal bones, and in some cases arrangements of stones where shelters probably once stood. Mary also identified some sediment levels at Olorgesailie as occupation sites where the members of this hand ax culture had camped.

Olorgesailie was filled with animal fossils and tools, but no remains of the toolmakers turned up.

Ape Island

By the late 1940s, Louis had spent more than twenty years looking for human ancestors in East Africa, searching in progressively older sediments. There was another approach to consider: start further back in time and try to work forward. In other words, look for fossil apes that might illuminate the split between the ape and human branches of the primate tree.

The Leakeys and other scientists had scouted Rusinga Island, on Lake Victoria, on several occasions. In 1932, Louis had found many mammals and the teeth, jaws, and limb bones of a number of apes that he had dubbed *Proconsul*. The place was littered with animal fossils from the early Miocene, the geological epoch that opened about 23 million years ago. There was a widespread bias that humans were so different from the great apes that the split had to be pretty deep in time in the Miocene. Louis had found a great deal of material in many short visits to Rusinga, and he imagined he could do much better if the place was properly excavated.

In the late 1940s, a joint expedition was opened with British colleagues to exploit the riches of Rusinga. The whole Leakey family, including their children, Jonathan and Richard, went along. Louis soon found a crocodile skull and started excavating it. Mary, who "never cared in the least for crocodiles, living or fossil," left Louis to his croc and went looking for something more interesting.

It was not long before she saw some promising bone fragments lying exposed, with a tooth protruding from a slope. "Could it be?" she wondered. She called to Louis, and as they brushed away the loose sediments from the tooth, a jaw emerged. Even better, it was clear that they had a good portion of the face, something that had never been recovered before.

In fact, this *Proconsul* skull was the first skull of a fossil ape of any age to be discovered. It was a great prize, but first it had to be put together. Mary worked long hours to piece together its more than thirty separate pieces, some as small as a matchhead. She and Louis were the first to see a *Proconsul* face. Louis thought that parts of the face looked human-like and leapt to the conclusion that *Proconsul* was a hominid and that humans' earliest ancestors were present in the Miocene.

Exhilarated, they wanted to celebrate and decided that the best way was to have another baby. Mary explained in her autobiography that "that night we cast aside care and that is how Philip Leakey . . . came to join our family" nine months later (Figure 11.4).

As Louis spread the news of their find, their British colleagues were eager to see it. Since Mary had found the skull, Louis thought she should take it to England, for there was some risk in sending such a precious original specimen from Nairobi to London. Louis worried that if it were lost or damaged, their chances of finding another one were slim.

Mary and *Proconsul* both got VIP treatment. BOAC (British Overseas Airways Corporation) offered her a free flight, and *Proconsul* traveled in a box on her lap. When Mary arrived at Heathrow Airport, the press was there en masse and asked her to pose on the gangway. She and the fossil ape were, much to her surprise, front-page news. Unaccustomed to such attention, Mary was relieved to hand

Figure 11.4 The Leakey Clan. A family portrait of (left to right) Richard, Mary, Philip, Louis, Jonathan, and the pack of family dalmatians. Photo from R. Leakey (1983), *One Life: An Autobiography* (Michael Joseph, London). Leakey Family Archives.

the skull over for further study at Oxford and return to Africa. As it turned out, some experts did not see it as a hominid but an early ape.

Rusinga became a favorite spot for the Leakeys' Christmas "holiday." Mary laid down bedding for the boys on top of the supply crates in the back of the family Dodge, and they would set out before sunrise for the Rift Valley and beyond. It was an all-day, four-hundred-mile-trip to Kisumu, on the shore of Lake Victoria, then an overnight boat ride to the island. Richard Leakey later recalled these adventures fondly:

> Those late night arrivals at Kisumu were always intensely exciting. The loading and the stowing had to be done by the boat lights. . . . The movement of the boat, the noise of the engines, and my father's commands all created an atmosphere of urgency and expectation in which we children reveled . . .
>
> I particularly enjoyed being awake on the boat in the early hours before dawn. As we made our way along the lake, a cool wind blew

and, periodically, a falling star punctuated the inky sky. By sunrise, usually glorious with warm red skies and golden clouds, we were well on our way. . . . We would start the day with marvelous fresh fish for breakfast. To taste pan-fried fish in the perfection of an African dawn must be one of the most delightful of all experiences.

The family lived on the boat and explored the island by day. The boys entertained themselves by fishing, playing with the local children, and occasionally finding fossils. At the age of six, Richard found his first—the complete jaw of an extinct pig. Unlike many other sites, there was fresh water to bathe in, if one did not mind the crocodiles. The family routine was for everyone to get ready to jump in, then Louis would fire a couple of shotgun blasts into the water which he believed ensured fifteen minutes of safety.

In the afternoon, Louis often took the boys prospecting for fossils. He pointed out birds and butterflies and showed them how to sneak up on wild game, to make tools out of stones, and to make fire by rubbing sticks together. The boys cherished these adventures, and they became keen naturalists. Rusinga yielded many unusual fossils that were exceptionally preserved as three-dimensional forms—including seeds, insects, small rodents, and even an ant colony—but *Proconsul* was the star.

While Mary had done her best to dodge the limelight, the *Proconsul* publicity had a very important tangible dividend beyond its scientific value: it brought attention and funding to the Leakeys' efforts that would enable them to sustain excavations, not just at Rusinga, but finally at Olduvai.

Dear Boy

For twenty years, Louis and Mary had scouted Olduvai but had not undertaken a proper excavation for two reasons. First, the gorge was very large. They had explored 180 miles of exposures that ranged from 50 to 300 feet in depth. It was important to be selective, and there was a large number of promising sites. The second was the shortage of funds and the necessity of doing other work to make ends meet.

The success of *Proconsul* attracted grants and benefactors. One of the latter was Charles Boise, a London businessman with a keen interest in prehistory. He had funded part of the Rusinga expedition; now he committed to supporting the Leakeys for seven years, so they decided to excavate at Olduvai, determined to find an early human.

The excavations went on throughout the 1950s and focused mostly on what was called Bed II, a lower site that yielded more than 11,000 artifacts and enormous numbers of large, exceptionally well preserved mammal fossils. They included, for example, complete skulls of a giant buffalo-like animal called *Pelorovis*, with horns spanning more than 6 feet. They also found a pile of *Pelorovis* bones that had been butchered by the tool-wielding inhabitants of that time. But the hominids themselves were ghosts: only two teeth were found in seven years.

In July 1959, the Leakeys turned their attention to Bed I, the oldest of the Olduvai beds. One morning Mary went out prospecting alone while Louis was sick in bed. She saw a lot of material lying at the surface, but one scrap of bone, projecting from beneath the surface, caught her eye. It looked like part of a skull—a hominid skull. She carefully brushed away a little of the sediment and saw parts of two large teeth in the upper jaw (Figure 11.5). They *were* hominid for sure, she thought. She had a hominid skull, and a lot of it, at that. Mary jumped into her Land Rover and raced back to camp, shouting, "I've got him! I've got him! I've got him!"

Louis asked, "Got what? Are you hurt?"

Mary blurted out, "Him, the man! Our man."

Louis recovered instantly from his illness and rushed to the site with Mary. He saw that it was indeed a hominid and that most of the skull was there. After twenty-eight years of searching, they had indeed found "their Man."

Their emotions were soon shared with and by their many colleagues in paleoanthropology.

First, however, they had to extract and piece together what Mary came to call her "Dear Boy." His skull was in some four hundred pieces. As she had with *Proconsul*, Mary patiently put him together. She had the upper jaw and teeth, most of the face, and the top and back of the skull. He was, in Louis' words, "quite lovely" (Figure 11.6).

Louis tried to decipher where this hominid fit in the scheme of human evolution. It bore some resemblance to the australopithecines that had been found in quarries and caves in South Africa. The first member of this group was described in the 1920s by Raymond Dart.

as an extinct ape with a brain larger than a chimpanzee—an intermediate between living apes and humans. Though none of the australopithecine skulls was as complete, Dear Boy's similarities to them posed a conundrum for Louis; he had seen the former as an evolutionary dead end, an offshoot of the line that gave rise to modern humans. Furthermore, no evidence of australopithecine toolmaking had been unearthed. Now he was looking at a skull that had come from tool-bearing sediments. The brain was too small to classify Dear Boy as a member of our genus *Homo;* it was smaller than Java Man and other *Homo erectus.* Yet he also noted so many differences between Dear Boy and the australopithecines that he decided it deserved a genus of its own. A new name also would help distinguish their discovery from prior fossils. He chose *Zinjanthropus* ("Man from East Africa") *boisei* (after their benefactor).

Louis zipped off a note to the journal *Nature,* describing the new hominid; he could not wait to spread the news and to show off his prize. He soon got the perfect stage—a meeting on African prehistory in South Africa. On the way to the conference, he stopped to see Phillip Tobias, a professor of anatomy and hominid expert. It was late at night when the Leakeys invited Tobias up to their hotel room and opened the padlocked wooden box holding "Zinj." Tobias was not prepared, for when he saw the fossil's face, "It absolutely sent shivers down my spine."

They then visited Raymond Dart, who knew how long and hard the Leakeys had searched. Dewy-eyed Dart told Louis and Mary, "I am so glad that this has happened to you of all people."

At the conference, the attendees were buzzing with the rumor that Louis had a big surprise for them. Louis could barely wait his turn. When he reached the podium, he raised Zinj up for all to see, and the audience broke into tremendous applause. Ever the showman, Louis did not bring a cast, a drawing, or a slide—he brought the real thing. Later, he invited his colleagues to gather by a table under a palm tree in a courtyard and allowed them to handle and inspect the specimen.

The press reaction was just as dramatic. Newspapers around the world announced the great discovery in headlines. Louis embarked on a speaking tour that included a triumphant night in London and sixty-six talks at seventeen universities across the United States. Audiences were captivated by the account of his long search and ultimate success in finding Zinj.

Figure 11.5 **Mary unearthing Dear Boy.** Top, Mary brushes away loose rock and sediment. Leakey Family Archives. Bottom, Dear Boy's palate emerges. Reprinted with the permission of Cambridge University Press.

Figure 11.6 Dear Boy. "Zinjanthropus," now known as *Australopithecus* or *Paranthropus boisei*. Photo by Javier Treveba/Photo Researchers.

The question of Zinj's age was on everybody's mind. How far back did our hominid ancestry reach? Louis told his audiences that he thought Zinj lived more than 600,000 years ago. This figure was based on a study of the Olduvai sediments and geologists' estimates of the age of the Pleistocene. But shortly after Zinj was discovered, two geophysicists used the new potassium-argon dating technique to obtain a figure of 1.75 + 0.25 *million* years for the ash bed above which Zinj had been found. The date was staggering—Zinj was three times older than Louis had thought (and which some believed to have been an exaggeration). The tools and the toolmakers were in fact far older than anyone had imagined.

The discovery and dating of Dear Boy changed the course of paleoanthropology, finally shifting everyone's undivided attention to

Africa, where Darwin and Louis thought it belonged. And they brought the time frame of human evolution into real terms.

They also changed the course of the Leakeys' life and work.

The National Geographic Society gave Louis the largest grant he had ever received, and the American public was so enthralled that they, too, began to support the research. The long drought for hominids, and funding, was over, and a whole new era of paleoanthropology was opening.

Destiny and dynasty

The next year at Olduvai, 1960, the excavation opened with a great sense of anticipation. Louis, now caught up in public relations and museum work, could only visit from time to time, but Mary, accompanied by her dalmatians, set up a permanent camp and led the team. A massive undertaking, the excavation was carried out on an unprecedented scale that involved more man—and woman—hours of labor in 1960 alone than in all of the nearly thirty previous years put together.

The children were now older. Jonathan, twenty, worked at Olduvai for a few months. One day he asked his mother, "Does any animal have a long thin bone like this?" as he traced a shape through the air with his finger. Mary said she couldn't think of one, and Jonathan casually said, "Oh, then I think it must be a hominid." Mary dropped her work and rushed to see what Jonathan was talking about. Sure enough, it was a hominid leg bone, a fibula. Later he found a tooth and toe bone. Mary decided to excavate "Jonny's site."

There, Jonathan found the remains of two individuals, including a skull. While just 100 yards from Zinj, the skull was about a foot below the level where Zinj had been found, so it was older. And it was different. It enclosed a larger brain, and its shape was more similar to that of a modern human. Amazingly, the excavation turned up twenty-one bones of a hand and twelve bones of a foot. It was no doubt a different species from Zinj, and Louis hoped this new find might be closer to our *Homo* line than Zinj.

He called for expert help, cabling Phillip Tobias: "Come quickly, top secret, we've got the Man." Various bones were wrapped and sent to experts in London. When one scientist opened the tins containing the foot bones, and fit them together, he later said his "hair stood on end. The foot was completely human." The hand bones evoked the same response. The tips of the fingers and thumbs indicated that this hand was quite capable of making the tools found nearby (Figure 11.7).

Several experts concurred that this new hominid belonged in our genus, and it was dubbed *Homo habilis*, meaning "handy, able, or skillful." Meanwhile, Zinj had been reclassified as a member of the *Australopithecus* genus. This meant there were two lines of hominids at Olduvai. The human tree was growing—and branching out.

Louis was not completely left out of the fossil-hunting picture at Olduvai. Late in the 1960 season, he, eleven-year-old Philip, and a geologist went out prospecting in the gorge. Louis spotted what he first thought was the shell of a tortoise. In fact, it was a hominid skull. He fetched Mary and beamed over his prize. Excavation revealed that this skull was different from either Dear Boy or *Homo habilis;* dating revealed it to be younger, at 1.4 million years old. It did have a strong resemblance to *Homo erectus* (*Pithecanthropus*) from Java. Louis had now found in Africa a much older *Homo erectus.* Three different hominids had lived at Olduvai over the span of a few hundred thousand years.

In the ensuing years, while the search for hominid remains and cultural artifacts remained a priority, Mary's meticulous, systematic approach to excavating throughout Olduvai and mapping every object produced a detailed record of the nearly 2-million-year history of Olduvai's habitat and animals, as well as hominid occupation sites.

Sitting on mountains of the oldest tools and a collection of the oldest known hominids, the Leakeys sat on top of the world of paleoanthropology. Before the decade was out, they would be joined by someone who would have been considered the least likely of all potential successors, their son Richard.

While Richard was a part of the family excavations as he grew up, he was determined to avoid paleontology, and he struck out on his own to establish a safari business. Extremely able in the bush, he found a pilot's license more useful than a university degree.

But he could not escape his roots. Flying tourists along Lake Natron one day in 1964, he spotted formations that looked like those at Olduvai and told his father about them. Louis supported a small expedition there, led by Richard, which turned up an *Australopithecus* specimen. Richard was hooked.

Figure 11.7 Hand of *Homo habilis*. The bones reveal that this hand had the precision to grip necessary for the fine manipulation of objects. Photo from *Human Origins: Louis Leakey and the East African Evidence* (1976), edited by G. I. Issac and E. R. McCown (W. A. Benjamin, Menlo Park, Calif.). Professor Michael Day, London University.

A few years later, Louis, in declining health and no longer able to lead expeditions, arranged for Richard to lead an expedition into Ethiopia's Omo Valley. This rough, beautiful country challenged all of Richard's skills. One day he had to beach a boat to escape a crocodile that had attacked and latched onto it. His temerity was rewarded with an early *Homo sapiens* skull; at about 130,000 years old, it was the earliest representative of our species known at the time. Later, on a flight back to camp, a thunderstorm forced Richard to detour over the eastern shore of Lake Rudolf (now Lake Turkana) for the first time. He again spotted promising rockbeds, which he soon determined were older than those in the Omo Valley, and he decided to explore them the next season.

One of his first finds was a remarkably complete skull of *Australopithecus boisei*, a perfect companion for his mother's Dear Boy. The fossils kept coming—including various parts of other *Australopithecus boisei*, *Homo habilis*, *Homo erectus*, and perhaps a fourth hominid—forty-nine specimens in all. The haul was larger than his parents' spoils from Olduvai. Though just in his late twenties, Richard had become the new star of paleoanthropology. In late September 1972, he flew to Nairobi to show his father a nearly complete, 1.9-million-year-old skull known then only as "1470." Louis was delighted by what appeared to be the oldest *Homo* specimen found, and father and son enjoyed one of their best moments together. Just five days later, Louis died of a heart attack while starting out on a lecture tour.

Richard's expeditions, too, became a family operation. He married Meave Epps, a member of his team, and in later years their daughters joined the excavations and were there when the first nearly complete *Homo erectus* skeleton, "Turkana Boy," was unearthed. Meave Leakey has since continued to extend hominid history further back in time and to redraw our picture of the human family tree.

But it was not just Richard and Meave who ensured that the Leakey name would remain synonymous with the quest for human origins, for Mary's adventures were far from over.

Something for the mantelpiece

Mary's years at Olduvai yielded tens of thousands of artifacts, including the oldest known tools and a diverse array of implements—choppers, chisels, cleavers, scrapers, picks, and more—of varying sizes and shapes. As early as 2 million years ago, hominids were fashioning specific tools for specific purposes. The diversity of hominids and tools found at Olduvai led to the inescapable deduction that there must have been yet older cultures in East Africa. However, Mary's excavations had reached the bottom of the Olduvai beds. Clues to those older cultures must lie elsewhere.

Mary had first visited Laetoli, some thirty miles from Olduvai, with Louis on their first trip in 1931. She made a few brief searches there over the decades until, in 1974, a hominid jaw and some teeth were found. When she realized that the fossil beds lay under 2.4-million-year-old ash, it was clear the fossils were considerably older than anything at Olduvai. Mary moved her operation to Laetoli.

The camp drew a lot of visitors. One day in 1976, three visiting scientists—Jonah Western, Kaye Behrensmayer, and Andrew Hill—got into a playful elephant dung–throwing fight. Falling, Hill discovered he had landed on a hard surface that appeared to display ancient animal footprints, and excavation revealed thousands of animal footprints of remarkable clarity. Apparently, a nearby volcanic eruption had been followed by a rainfall that cemented the fresh prints of many creatures in place. They were quickly covered by more ash and left undisturbed for 3.5 million years.

Many more footprint exposures were located, and Mary made their documentation a priority. In 1978, however, some prints were uncovered that were not animal but decidedly hominid. Excavation revealed two parallel tracks of hominid footprints extending for about 80 feet. One set was smaller than the other, indicating that a juvenile or female was beside an older or male individual. (Detailed examination of the larger tracks suggested that the larger hominid was either shuffling his feet or that another youngster was walking in the tracks of the larger one.) Skeletons and leg and foot bones, which can indicate whether an individual walked upright or not, were and are very hard to come by for ancient hominids—and none had been found at Laetoli. There, Mary stared at the most vivid evidence imaginable of the bipedal gait of our ancestors 3.5 million years ago (Figure 11.8).

Africa had saved the best for last. After excavating a particularly sharp set of prints herself, Mary lit a cigar and, admiring the impressions, declared, "Now this is really something to put on the mantelpiece."

Figure 11.8 **Something for the mantelpiece.** Mary admires a freshly excavated footprint at Laetoli. Photo courtesy of John Reader, from J. Reader (1981), *Missing Links: The Hunt for Earliest Man* (Little, Brown and Company, Boston).

Louis and Mary Leakey had lifted the veil over human origins and revealed that upright-walking and toolmaking hominids had long preceded modern humans. But, as with all science, their successes gave rise to a new set of questions. Just how far back in time did hominid history go? When did our line split off from that of the great apes? When and where did our species, *Homo sapiens*, evolve? How are we related to other humans, such as Neanderthals? Answers to such questions would come, to the surprise of many, not so much from stones and bones but from an altogether new science and a different breed of scientists than the intrepid explorers of the Rift Valley. The epicenter of this new revolution in paleoanthropology was half a world away, in a couple of laboratories in California.

Figure 12.1 Ava Helen and Linus Pauling. On the coast at Corona Del Mar in 1924; Linus was twenty-three. Their marriage was very close and lasted 58 years. It was Ava Helen who inspired Linus' later activism. From the Ava Helen and Linus Pauling Papers, Special Collections, Oregon State University.

❧ 12 ❧
Clocks, Trees, and H-Bombs

One man with courage makes a majority.
—*Andrew Jackson*

HE WAS WITHOUT QUESTION the greatest chemist of the twentieth century. He pioneered a new understanding of the nature of chemical bonds. He discovered one of the keys to the complex structures of proteins. He deciphered that sickle cell anemia was due to abnormal hemoglobin, the first demonstration of the molecular basis for a human disease. His discoveries earned him his first Nobel Prize, for Chemistry, in 1954. His brave leadership of the movement to ban the atmospheric testing of nuclear weapons earned him his second Nobel Prize, for Peace, in 1963—making him the only person ever to have won two unshared Nobels.

As an evolutionary scientist, however, Linus Pauling was a late bloomer. While overshadowed by his many other accomplishments, Pauling, late in his career and during an extremely difficult period due to the controversies in which he was embroiled, was a pioneer in a new field of evolutionary biology. The adventures that led to his major insight took place not in a tropical jungle but in the political jungle of the Cold War. It was a circuitous journey, but one well worth retracing.

A chemist with a conscience

Pauling was trained as a chemist and physicist. Early in his career, he focused on developing a set of rules that explained the formation of chemical bonds. In over a decade of work begun in the latter half of the 1920s, Pauling transformed chemistry from a field that had relied mainly on observation to a more predictive science, grounded in physical principles and centered on chemical structures. His efforts culminated in a landmark book, *The Nature of the Chemical Bond* (1939).

As World War II broke out, Pauling offered his services, and his laboratory at Caltech, to the U.S. government. He worked on new explosives and propellants, developed a meter to monitor oxygen levels in pressurized spaces such as submarines and airplanes, and invented a synthetic blood plasma for battlefield transfusions. He was given awards by the navy and the War Department for these accomplishments and, in 1948, President Truman awarded him the Presidential Medal of Merit, the highest civilian honor, for "extraordinary fidelity and exceptionally meritorious conduct," and cited him for turning his "imaginative mind to research on military problems with brilliant success."

Pauling was often asked to join committees of scientists to study one issue or another. Shortly after the war, he was asked to join the Emergency Committee of Atomic Scientists, chaired by Albert Einstein. Alarmed by the power and potential spread of atomic bombs, the group was dedicated to informing the public about the dangers of nuclear weapons. Pauling's wife, Ava Helen (Figure 12.1), had a deep commitment to issues concerning peace, social justice, and human rights. She encouraged him to get involved in what she called "peace work" by persuading him that his scientific work would not mean very much if the world were destroyed. As it turned out, this new work inspired his interest in evolution.

Pauling used his many speaking opportunities to call for negotiations to end the emerging arms race between the United States and the Soviet Union. But his outspoken views drew the attention of U.S. government officials, who were concerned with flushing out communists and communist sympathizers in any position of prominence.

Despite his exemplary work in the war effort and a public declaration that he was a "Roosevelt Democrat" and had no affiliation with or any kind of sympathy for communism or communists, suspicions fell on Pauling. The FBI started keeping close tabs on what he said or wrote and who his associates were. In 1952, he was denied a passport to attend a critical scientific conference in London that was being convened to discuss his revolutionary work on protein structure. The State Department's refusal to grant him the passport led to an international furor over the stifling of one of the country's most accomplished and prominent scientists. It might have also affected Pauling's shot at solving the structure of DNA, for he missed the opportunity to see Rosalind Franklin's critical new X-ray crystallographic pictures of DNA. Pauling, the leading authority on the structure of large biomolecules, was also working, albeit slowly, on DNA. James Watson and Francis Crick used the X-rays to help crack the structure within the next year.

Pauling's political troubles continued. In late 1952, an FBI informer testified in front of a House committee that Pauling was a "concealed Communist." In late 1953, his research grants from the Public Health Service were canceled, and new grant applications were ignored. His application for a passport for a pleasure trip to India was also denied. Pauling was getting the message to keep quiet, and he restrained himself from antagonizing the authorities—for a while.

Two bombshells

On March 1, 1954, the United States dropped a bomb on the Pacific island known as Bikini Atoll, obliterating it. It was supposed to be a secret test, but the blast was much larger than the scientists had expected: the equivalent of 15 megatons of TNT was released instead of the 4–8 megatons expected. The wind also changed direction from the forecast, sending radioactive debris over populated islands instead of the open sea. A fine ash fell on a Japanese fishing boat trawling ninety miles from Bikini. By the time it reached port, the crew was suffering badly from radiation sickness; one member died. It soon became clear that the Bikini blast was something new—a "superbomb."

It was an H-bomb. The new design was the brainchild of the physicist Edward Teller, a committed hawk who would have no part of Einstein's and Pauling's efforts to halt the arms race. The bomb was in fact the largest ever detonated by the United States, packing a thousand times the power of those dropped on Hiroshima and Nagasaki.

Pauling was horrified. The arms race was escalating and the dangers were increasing. The superbomb blew radioactive material high into the atmosphere, where it was carried around the planet and returned as fallout. That material included new, exotic isotopes that had not been detected in previous blasts. The government's prior claim that the growth in the strength of the atomic bomb was not accompanied by a growth in the amount of radiation released was clearly dubious. Pauling again voiced his concerns.

The authorities were watching him, and on October 1, 1954, his application for a passport was once more denied. But on November 3 the tide turned. Just before giving a lecture at Cornell University, he got a phone call from a reporter. "What is your reaction to winning the Nobel Prize in Chemistry?" he asked, taking Pauling completely by surprise.

He was delighted, of course. While most Nobel prizes recognize a particular discovery, Pauling's citation encompassed three decades of work. After he hung up, he entered the lecture hall and was greeted by a standing ovation.

But he soon wondered if he would be allowed to go to Stockholm to claim his prize. The Ambassador to Sweden wondered, too, and cautioned Secretary of State John Foster Dulles: "I must emphasize that if passport is refused effect on Swedish public opinion of all shades will be catastrophic."

The communist-hunting bureaucrats and the FBI did not like it, but Pauling got his passport, not just for Stockholm, but for the world. And he took advantage of it, embarking on a five-month worldwide tour during which he was feted and praised. He also heard concerns across the globe about the new era of making and testing superbombs, and he felt the anxiety people had that anyone would consider using such weapons. Returning to the United States, he was determined to use his talents and his prestige to oppose the arms race.

Pauling tried to learn everything he could about the new bomb. That was difficult, as the government was reluctant to divulge design details or test data. Researchers around the world, however, were focusing on the radioactive fallout and discovering some unsettling facts. They found, for example, that the Bikini blast produced strontium-90, an isotope never before seen on the planet and one that could reach the food chain and expose millions to radiation.

Pauling brought himself up to date on radiation's effects on the body, on genetics and mutation rates in particular. Fortunately, he had world experts in genetics and radiation, including the future Nobel Laureate Edward Lewis (1995), right next door at Caltech. Pauling extrapolated from animal studies to consider the effect of increased radiation in the atmosphere on birth defects, miscarriages, and overall human health. He took on government experts, including Edward Teller, "the father of the H-bomb," who downplayed the health risks of radiation when compared to the importance of nuclear weapons to national security.

Pauling drew on his studies of sickle cell anemia to make the public understand the link between mutated genes and disease. He also rallied and enlisted the help of his fellow scientists, who understood the dangers and also felt they were being downplayed by the government, to press for a ban on testing nuclear weapons. In the spring of 1957, Pauling circulated a petition among scientists against nuclear testing. The petition included the statement: "As scientists we have knowledge of the dangers involved and therefore have a special responsibility to make those dangers known." The petition eventually gathered more than eleven thousand signatures from forty-nine countries. Pauling and his wife personally submitted it to the United Nations in January 1958.

All of this time, Pauling was living a double life, although not the one alleged by the FBI and the State Department. He had become a highly visible activist, but he was still a working scientist. He spent part of his time in his laboratory in Pasadena, directing its studies of protein structure, the other part giving speeches, writing articles, lobbying for the test ban, and debating his opponents. In order to counter arguments such as those made by Teller, who suggested in a televised debate that it was even possible that a bit more radiation in the atmosphere could have a positive effect on evolution, Pauling had to catch up on genetics, mutation, and evolutionary theory. In early 1959, he bought a copy of Darwin's *Origin of Species* and George Gaylord Simpson's *Meaning of Evolution*. The chemist was evolving into a biologist.

By late 1959, when the world was just learning about the Leakeys' discovery of *Zinjanthropus*, Pauling's homework on nuclear weapons had him thinking a lot about evolution. And that is also when Emile

Zuckerkandl, a young researcher from France, joined Pauling's laboratory. Zuckerkandl quickly sparked a connection between Pauling's two lives and the birth of a new way of studying evolution and human history.

A molecular clock

In the fall of 1959, molecular biology was still a toddler. The genetic code—the means by which a sequence of bases in DNA determines the sequence of amino acids in a protein—was not yet known (for a short primer on DNA and proteins, see the box on page 252). The methods for determining the sequences of proteins were difficult. Few sequences were known, let alone sequences from different species. It was generally understood that mutations in DNA would cause changes in proteins and that this must be part of evolution. But no one knew how similar the proteins of different species were. This mystery is what Pauling thought Zuckerkandl should try to solve.

Because hemoglobins were the best-studied proteins, it made sense to start with them. Zuckerkandl collected the hemoglobins from gorillas, chimpanzees, orangutans, rhesus monkeys, cows, pigs, and an assortment of fish. It was not feasible to determine the exact sequences of amino acids in the two chains, called alpha ("α") and beta ("β"), of the hemoglobin protein; it would take a small army of laboratories to decipher them. But a "quick and dirty" technique was available; it involved "fingerprinting" proteins by digesting them with an enzyme and examining the pattern of the fragments produced. Zuckerkandl found that the human, gorilla, and chimpanzee patterns were nearly identical, the orangutan just a little different, and the cow and the pig much more so. This was some evidence that proteins might reflect evolutionary relationships to a degree, but nothing more could be said without a more direct comparison of protein structure.

Zuckerkandl then teamed up with Walter Schroeder, a Caltech protein chemist, to decipher the numbers (not the order) of each amino acid in the gorilla α- and β-chains. Their analysis indicated that the human and gorilla α-chains differed by just 2 out of 141 total amino acids, the β-chains by 1 out of 146 amino acids. The proteins were indeed very similar; the gorilla protein was no more different from the human protein than were known human variants such as the sickle cell hemoglobin.

Because of his prominence, Pauling was often invited to conferences all over the world or "Festschrifts" to honor senior scientists. These gatherings were often marked by the publication of a book with the participants' papers. Pauling asked Zuckerkandl to write a paper with him in which "we should say something outrageous!" Because these would not be peer-reviewed, like normal journal articles, they had considerable liberty to say anything they wanted.

By the time they were writing the paper, much more information had appeared on the composition and sequence of mammalian hemoglobins. It occurred to Pauling and Zuckerkandl that there might be some interesting evolutionary history to be gleaned. They noted that the number of differences between α-chains or β-chains of different species increased in accordance with evolutionary distance. They realized that if they plotted "the number of differences between corresponding chains in different animal species, and the geological age at which the common ancestor of the different species in question may be considered to have lived," they could estimate the average number of years that elapsed per substitution in a globin chain. Using the estimated divergence time of horses and humans (130 million years) and the number of differences between their α-chains, they calculated a figure of about 14.5 million years per change. Given the data, anyone might have made that observation.

But then Zuckerkandl and Pauling took a leap. They used this "molecular clock" of one change in a globin chain per 14.5 million years to estimate the age of other ancestors based simply on the number of differences in their descendants' hemoglobin chains. Most interesting was their analysis of the gorilla and human. These species appeared to bear just two differences in their α-chains (later found to be one) and only one difference in their β-chains. The two scientists figured that the one or two changes between the gorilla and human globin chains translated to their last common ancestor's having lived 7.3–14.5 million years ago. (To get these numbers, they multiplied the number of changes by 14.5 and then divided by 2, because the changes occurred across the two lineages.) They settled on the average estimate of these two numbers—11 million years—and noted that their calculation "falls on the lower limit of the range estimated on paleontological grounds," which then ranged from 11 to 35 million years.

It was a simple, elegant, and revolutionary idea. If the sequences of biological molecules could be used to peer back into the mists of time, they would indeed harbor unique information about evolutionary history.

It was a simple idea all right, and, not surprisingly, some thought it was too simple to be true.

Doubt

Two lions of evolutionary biology, Ernst Mayr and George Gaylord Simpson, were among the most prominent and vocal skeptics. Mayr, a renowned expert on species and systematics, and Simpson, a leading paleontologist, were two of the architects of the so-called Modern Synthesis of evolutionary theory that emerged in the 1940s. The Modern Synthesis united genetics, paleontology, and systematics into a harmonious theory that asserted that the variations and small changes observable within populations and species could explain the larger differences that evolved between species and higher taxonomic ranks over long periods of time.

The difficulty that Mayr and Simpson had with the "globin clock" was that it assumed that molecules accumulated differences at a regular rate, simply as a function of evolutionary time. But both men knew from natural history and the fossil record that the visible rate of evolution varied greatly. Sometimes evolutionary change proceeded quickly, sometimes species remained the same for long periods. Mayr, for example, pointed out that humans' evolution into a "bipedal, tool-making, speech-using hominid necessitated a drastic reconstruction of his morphology, but his morphology did not, in turn require a revamping of his biochemical system. Different characters . . . thus diverged at different rates." Mayr expected that different molecules would also change at different rates and therefore give different answers to questions about the timing of past events.

At scientific meetings and in print, Zuckerkandl and Pauling and their critics volleyed back and forth. Pauling loved the idea of the clock and its potential to study a wide range of time periods. He was optimistic that "it will be possible, through the detailed determination of amino acid sequences . . . to obtain much information about the course of the evolutionary process."

But in other, more global matters, he had good reasons to be pessimistic.

Crisis

In 1962, after a short pause, President Kennedy decided to resume testing nuclear weapons. Pauling, who was fond of Kennedy but uncompromising on the matter of arms testing and war, sent him a telegram:

> *March 1, 1962*
>
> *President John F. Kennedy, White House:*
>
> *Are you going to give an order that will cause you to go down in history as one of the most immoral men of all time and one of the greatest enemies of the human race?*
>
> *In a letter to the New York Times I state that nuclear tests duplicating the Soviet 1961 tests would seriously damage over 20 million unborn children, including those caused to have gross physical or mental defect and also the stillbirths and embryonic, neonatal, and childhood deaths from the radioactive fission products and carbon 14.*
>
> *Are you going to be guilty of this monstrous immorality, matching that of the Soviet leaders, for the political purpose of increasing the still imposing lead of the United States over the Soviet Union in nuclear weapons technology?*
>
> *Linus Pauling*

Despite this message, Pauling was invited to a dinner party at the White House for American Nobel Prize winners on April 29, 1962. Pauling and his wife spent the days of April 28 and 29 with other demonstrators against the resumption of weapons tests by picketing outside the White House (Figure 12.2). Then, on the evening of the twenty-ninth, they changed into formal attire, joined the dinner party, and even got up to dance.

Later that year, in October, U.S.-Soviet tensions escalated when the United States discovered that nuclear missile bases were being erected in Cuba. Kennedy and Premier Nikita Khrushchev's showdown during the Cuban Missile Crisis brought the two superpowers to the brink of nuclear war, and perhaps to their senses.

Once the crisis was resolved, both countries realized that they needed a treaty to stem the escalating arms race. In July 1963, both parties signed the Nuclear Test Ban Treaty.

On October 11, the day after the treaty went into effect, the Paulings were relaxing with friends at their cabin at Big Sur, on the California coast. With no telephone, it was one place where they could get some respite from the press, or so they thought. There was a knock on the door. A forest ranger had come to say that their daughter Linda was trying to reach them. After walking a mile to the ranger station, Pauling called Linda to hear her ask: "Daddy, have you heard the news?"

"No, what news?" Pauling replied.

"You've been awarded the Nobel Peace Prize!"

Molecules as documents of history

Pauling received, however, a mixed reaction to his prize. His chemistry colleagues and the president of Caltech yawned. His activities and frequent absences were sore points. Several members of Caltech's Board of Trustees, especially those with ties to the defense industry, had long wanted to get rid of Pauling. The Caltech president had, over the years, pushed Pauling out of his chairmanship of the chemistry division and reduced his laboratory space.

Miffed that the university essentially ignored his second Nobel Prize, Pauling called a press conference and shocked everyone at the university by announcing that, after forty years, he was resigning from Caltech. He was gone by the end of the year, moving to a think tank in Santa Barbara.

Zuckerkandl then shouldered most of the weight of the molecular clock problem. As more data became available, he had more with which to test the idea. Comparative study of another protein called cytochrome c, found in a wide range of organisms including animals, plants, and fungi, revealed that substitutions also occurred in the protein as a function of overall time. Zuckerkandl noted that the total number of substitutions per unit of time was not the same in globins and cytochrome c. In the latter, about half of all sites never changed even over the vast evolutionary distance separating yeast and humans, while in globins most sites were seen to vary among some species. Nevertheless, the fraction of changes that occurred at variable positions in each protein within a given time period was similar. Both molecules were keeping time.

Zuckerkandl and Pauling wrestled with the patterns of substitutions taking place in individual proteins. They understood that, even if their sequences had changed, globins and cytochrome c had the

Figure 12.2 Pauling protesting outside the White House. Pauling carried this placard during the day, then in the evening joined President Kennedy for a special White House dinner honoring American Nobel laureates. Photograph courtesy of United Press International.

same biochemical jobs to do, whether in a fish or a human. As biochemists, they knew very well that there were just a few categories of amino acids—some were positively charged, some negatively charged, some had no net charge. They reasoned that certain substitutions in proteins, of one amino acid for another with similar properties, would probably have little or no effect on protein activity; they were functionally "neutral" or "nearly neutral."

This insight was crucial to explaining the apparent constant rate of molecular change with the preservation of molecular function. If certain sites were allowed to change in proteins because they had little functional effect, then the steady beat of mutation in DNA would be manifest as a steady rate of change in protein sequences over time.

But this thinking was almost heretical to organismal biologists at the time, for they viewed all evolutionary change as brought about by natural selection and adaptation. George Gaylord Simpson, in a prominent critique of the molecular clock idea, wrote: "There is a strong consensus that completely neutral genes or alleles must be very rare if they exist at all. To an evolutionary biologist it therefore seems highly improbable that proteins . . . should change in a regular but non-adaptive way."

Zuckerkandl and Pauling argued that the similarity or differences in the "looks" of two organisms need not be reflected at the level of proteins—that visible change and molecular change were not necessarily coupled. From their biochemical perspective, they did not see a reason that the two must be linked. One or a few changes could lead to large differences, but many changes could also occur without causing functional change. They concluded that "changes that occur at a fairly regular over-all rate would be expected to be those that change the functional properties of the molecule very little . . . *There may thus exist a molecular evolutionary clock.*"

Zuckerkandl and Pauling had offered a new picture of evolution that was invisible to paleontologists and taxonomists—a picture of molecules ticking off evolutionary time without affecting how organisms looked, behaved, or functioned.

By 1966, a new breed of molecular biologists was excited by the possibilities of using molecules to document evolutionary history—to determine species relationships and to look back in time. Paleontologists and organismal biologists were still doubtful, and their skepticism soon erupted into white-hot controversy.

Shaking the hominid tree

One of the scientists who was particularly intrigued by Zuckerkandl and Pauling's ideas was Allan Wilson. Raised on a New Zealand cattle ranch, a novelty in a land where sheep outnumber humans by more than ten to one, Wilson went to the United States to pursue his Ph.D. in biochemistry. His family hoped that he would be gone for only several years. Instead, he stayed for his entire career and became one of the most innovative and influential figures in the new science of molecular evolutionary biology.

As a new assistant professor at Berkeley, Wilson focused on what molecules could reveal about primate evolution, including human origins. He was always interested in pushing new techniques and had developed an expertise in a highly sensitive way of analyzing protein relationships. Rather than requiring full protein sequences, which were difficult to obtain, it used antibodies to detect similarities and differences among proteins. The principle was simple: antibodies are produced in response to foreign substances—in this case, the injection of human serum albumin into a rabbit—and they bind to specific places all over the protein. If other albumin proteins—say, from a chimpanzee or some monkey—are different, the antibodies bind less well in proportion to these differences. The great advantage of Wilson's technique was that it was fast, quantitative, and could be used on proteins from any source without knowing their structure beforehand.

Wilson and Vincent Sarich, a graduate student in anthropology, used the antibody test to compare the albumins of a wide range of primates. They were pleased to find that the results reflected what was then currently thought about the relationships between primates. They found that, compared to human albumin, chimpanzee and gorilla albumins were the most similar, followed by that of Asiatic apes (gibbon, orangutan, and siamang), Old World monkeys, New World monkeys, and prosimians (e.g., lemurs and tarsier), respectively.

Their results were not only consistent with the primate evolutionary tree, but in order to be so, Sarich and Wilson concluded that the albumin molecule evolved at a steady rate, with any two species' sequences growing proportionally more dissimilar with time. If albumin evolution was keeping time, they realized they could use it to calibrate the primate tree, including the timing of the origin of the human branch. Paleontologists had generally pegged that at about 20 to 30 million years ago (Figure 12.3, top).

However, when they used their molecular tests to determine the time scale of hominid evolution, they got a startlingly different figure. They first calibrated their clock by estimating the age of one branch of the tree from the fossil record. They reasoned from the available, albeit fragmentary, evidence that the split between apes and Old World monkeys occurred roughly 30 million years ago. Then they determined from their antibody tests that the differences between Old World monkey albumins and human albumin was about six times greater than that between chimp or gorilla albumin and human albumin. They then reasoned that the human and the chimp and gorilla lines must have split more recently, just one sixth of the 30 million years, or 5 million years ago (Figure 12.3, bottom).

This was a straightforward calculation, but it led to a stunning conclusion—humans and chimps and gorillas shared a much more recent common ancestor than anyone had thought. If the human line split off from the chimpanzee line 5 million years ago, those 2-million-year-old fossils coming out of Olduvai Gorge represented a much larger fraction of overall hominid evolutionary history than the Leakeys thought. Sarich and Wilson's conclusion was also remarkably bold in that it came from a field still in its infancy. Relatively few scientists were familiar with the methods used or with the clock theory. And because these tools were at odds with more familiar fossil data, many were inclined to dismiss Sarich and Wilson's findings as fantasy.

Not the least of whom was Louis Leakey himself. After Wilson and Sarich published a second report, in which they confirmed their findings using other molecules and other fossils to calibrate their clock, Leakey wrote a blistering critique. He stated that Wilson and Sarich's conclusions were "entirely contrary to the most recent paleontological evidence" and based upon "a serious fallacy" and "an unsupported assumption for which there is no paleontological justification at the present time."

Leakey reviewed the paleontological evidence against Sarich and Wilson. He explained that the Kenyan fossils known as *Kenyapithecus wickeri* were hominids that had been reliably dated at 12–14 million years old. Moreover, these fossils were *"closely related"* to another Asian hominid, *Ramapithecus*, of about the same age. Because distinct ape fossils were known from the same locale in Kenya, Leakey concluded that hominids "were fully distinct from apes . . . about 12 to 14 million years ago. The date of separation suggested by Wilson and Sarich . . . is not in accord with the facts available today." He also rejected Wilson and Sarich's inference regarding the origin of Old World monkeys 30 million years ago.

The chorus of critics included Elwyn Simons, a prominent Yale paleontologist and champion of *Ramapithecus*, who chided the interlopers: "Students of human origins will know, however, that the story of hominid origins begins much earlier than this [Sarich and Wilson's date], since hominids of the genus *Ramapithecus* date back to ... about 14 million years ago." Wilson and Sarich had concluded that the last common ancestor of great apes and humans was also very recent, perhaps 7–10 million years ago. Simons confidently rejected that notion: "A common ancestor for all the living hominoids [great apes and humans] *could not have been living* as late as 7 million years ago ... or 10 million years ... but much more probably existed only prior to about 35 million years ago."

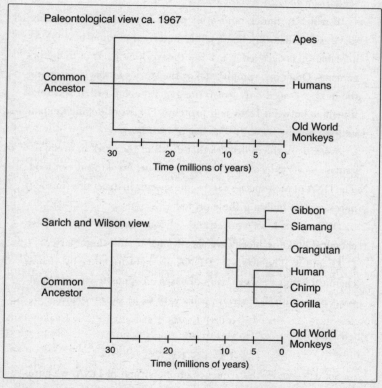

Figure 12.3 A new time scale for hominid evolution. Top, the pale-ontological view of primate and human evolution circa 1967. Bottom, Sarich and Wilson's albumin clock indicated a much more recent origin of the great apes (10 million years) and of the splitting of the lineage from the great apes (5 million years). Based on data in V. M. Sarich and A. C. Wilson (1967), *Science* 158: 1200–1203, drawing by Leanne Olds.

B IOLOGISTS COMPARE THE DNA AND PROTEIN SE-
QUENCES OF SPECIES AT MANY DIFFERENT SCALES,
FROM VERY CLOSE RELATIVES TO VASTLY DIFFERENT LIFE
forms whose lines split off from one another early in life's history.
*The clues to evolution come from understanding the meaning of the
similarities and differences we find.*

In order to understand a bit about how DNA and proteins
serve as documents of history, it is important to know how DNA's
information is decoded in making the working parts of living organisms. Don't be intimidated. For the stories in this chapter and
the next, we need only grasp the general composition and relationships between DNA and proteins. For more detailed explanations, see my book *The Making of the Fittest.*

Proteins are the molecules that do all of the work in every or-
ganism—carrying oxygen, building tissue, breaking down food.
The DNA of each species carries the specific instructions (in code)
necessary for building these proteins.

DNA is made of two strands of four distinct bases. These
chemical building blocks are abbreviated by the single letters A,
C, G, and T. The strands of DNA are held together by strong
chemical bonds between pairs of bases on opposite strands: A always pairs with T, C always pairs with G, as shown below:

~ A G T C A G T C ~
| | | | | | | |
~ T C A G T C A G ~

So, if we know the sequence of one strand of DNA, we auto-
matically know the sequence of the other strand. It is the unique
order of bases in a sequence of DNA (ACGTTCGATAA, etc.) that

continued on page 254

John Buettner-Janusch, an anthropologist at Duke University, was also critical: "If Sarich and Wilson had looked more carefully at paleontological investigations, they would have found their suggestion is unwarranted. . . . I object to careless assumptions and thoughtless statements about evolutionary processes in some of the conclusions drawn from the immunological data." Moreover, Sarich and Wilson's approach was just too damn easy compared with the rigor and labor of fieldwork and painstaking reconstruction. Buettner-Janusch complained: "No fuss, no muss, no dishpan hands. Just throw some proteins into a laboratory apparatus, shake them up, and bingo!—we have an answer to questions that have puzzled us for three generations."

The battle lines had been drawn. Somebody was wrong. Each side was confident it was the other. The paleoanthropologists, relying on well-known and long-studied fossils such as *Ramapithecus*—which Leakey characterized as "overwhelming evidence" of early hominids —concluded that the molecular timekeepers had to be wrong. Simons suggested, "I am not a biochemist . . . but it would appear that the number of basic assumptions underlying the present immunological data is considerably greater than that which affects the uncertain interpretation of fossils."

Wilson and Sarich continued gathering data and testing the molecular clock. They pointed to the 100 percent identical sequences of human and chimpanzee globins and used an independently calibrated mammalian globin clock to calculate that the odds were 1,000 to 1 against the last human-chimp common ancestor being older than 15 million years. Sarich showed that the small differences they had measured between chimp and human albumins were similar to those for other closely related pairs of species, such as goats and sheep, dogs and foxes, and horses and donkeys. He found himself "utterly unable to interpret the protein data" on chimps and humans in any other way than a very recent split. He was so convinced, he boldly stated that "one no longer has the option of considering a fossil specimen older than about 8 million years as a hominid *no matter what it looks like.*" Telling paleontologists, who had made their reputations by asserting an older hominid ancestry, that they could not possibly be right was not at all well received.

The controversy raged for better than a decade. Wilson and Sarich's work was generally dismissed by the anthropological community. But the close molecular relationship of humans to chimps forms the unique instructions for building each protein. Mutations are changes that occur in the DNA sequence; they arise at random throughout DNA at a fairly steady but low rate. They accumulate over time so that any two populations or species differ in their DNA sequence roughly in proportion to the amount of time that they have been separated.

How are proteins built and how do proteins know what their job is? Proteins themselves are made up of building blocks called amino acids. Each amino acid is encoded as a combination of three bases or a triplet (ACT, GAA, etc.) in the DNA molecule. The chemical properties of these amino acids, when assembled into chains averaging about 400 amino acids in length, determine the unique activity of each protein. The length of DNA that codes for an individual protein is called a gene.

The relationship between the DNA code and the unique sequence of each protein is well understood because biologists cracked the genetic code forty years ago. The decoding of DNA in the making of proteins occurs in two steps. In the first step, the sequence of bases on one of the strands of the DNA molecule is transcribed into a single strand of what is called messenger RNA (mRNA) (Figure 12.4). In the second step, the messenger RNA is translated into the amino acids that build the protein. In the cell, the genetic code is read (from the RNA transcript) three bases at a time, with one amino acid determined by each triplet of bases; a short example is shown in the figure.

There are sixty-four different triplet combinations of A, C, G, and T in DNA but just twenty amino acids. Multiple triplets code for particular amino acids (and three triplets code for nothing,

continued on page 256

grew even more obvious. Wilson and his student Mary-Claire King showed, in another landmark paper in 1975, that many chimp and human proteins were identical—or at least so very similar that it was difficult to explain differences in their respective anatomy and behavior in terms of the few differences in their proteins. Wilson, by studying and comparing the rates of molecular and physical change in frogs, birds, and mammals, saw that the two were uncoupled. Great physical changes could evolve among molecularly similar and recently diverged species. He concluded that looks could be deceiving —so deceiving, in fact, that he thought the interpretation of anatomy was too subjective, too prone to observer bias or error to be reliable.

And this is exactly the nature of the mistakes the paleoanthropologists had made that had put them and Wilson at odds. It turns out that Leakey's "overwhelming evidence" of early hominids was overwhelmingly wrong. Paleoanthropologists had biases about characters they would find in early hominids that would distinguish them from apes. They included certain dental features, such as small canine teeth. Fragmentary parts of *Ramapithecus* fit that picture, and it was accepted as the oldest hominid. But similar anatomical features can evolve independently in different groups. The subsequent discovery of more complete specimens of a relative of *Ramapithecus*, called *Sivapithecus*, showed more details about the jaw, teeth, and face. Reconstructions of the face revealed features shared with the orangutan. The new specimens also had thick enamel, large molar teeth and a robust lower jawbone, which were previously thought to be diagnostic of hominids. But *Sivapithecus* was not a hominid, nor therefore was the great *Ramapithecus;* they were actually closely related to orangutans and not in the hominid line. The same fate befell *Kenyapithecus*, which Leakey, based on its teeth, had mistakenly identified as "a very early ancestor of man himself."

As the contradictory fossil evidence crumbled, after more than a decade and a half of wrangling, in 1982 the paleoanthropologists finally came around to accepting Sarich and Wilson's 5-million-year-old chimp–human ancestor. Louis Leakey had long since died, but his son Richard confessed his conversion at a meeting in London: "I am staggered to believe that as little as a year ago I made the statements that I made about the molecular clock data. . . . I think the molecular people are closer to the truth than we've ever given them credit for."

Wilson and Sarich's conclusions were, in fact, confirmed many marking the stopping point in the translation of RNA and the making of a protein—the way periods mark the end of a sentence). Certain sets of amino acids have very similar properties, which is what led Pauling and Zuckerkandl to realize that proteins could change in sequence without changing in function.

Given a specific DNA sequence, it is easy to decipher the protein sequence that DNA encodes, so the genome of any species can be readily decoded.

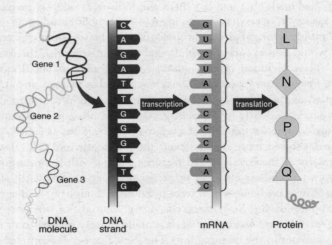

Figure 12.4 The expression and decoding of DNA information. This is an overview of the major steps in decoding DNA into a functional protein. Left, long DNA molecules contain many genes. The decoding of a portion of one gene is shown in two steps. First, the complement of one DNA strand is transcribed into mRNA. Then, the mRNA is translated into protein, with three bases of the mRNA encoding each amino acid of the protein (shown as the letters L, N, P, Q here). In mRNA, the base U is used in place of the T in DNA. From Figure 3.2 in S. B. Carroll (2006), *The Making of the Fittest* (W. W. Norton, New York).

times over with a variety of proteins and, as new technologies for DNA analysis and sequencing became available, with DNA clocks as well. With the advent of rapid DNA analysis, the molecular revolution foreseen and ushered in by Zuckerkandl and Pauling blossomed. Because molecules could circumvent the biases inherent in human classification schemes, biologists began using molecular information to build all sorts of evolutionary trees and to date past events.

Accepting what the "molecular people" had to say about the time scale of hominid evolution was not necessarily bad news for paleoanthropology. That 5-million-year figure for the last common ancestor of chimps and humans put an upper limit on the age of fossils that could be on the human branch of the evolutionary tree. The good news for paleontologists was that, by the early 1980s, they already had *Australopithecus afarensis* fossils that were about 3.5 million years old. The fossil hunters were much closer to the origin of the human branch, just 1 or 2 million years away, not the 20 million or so that they had once believed. By targeting deposits that were 4–5 million years old, they could zero in on the first stages of human origins.

The revolution in "molecular anthropology" was not over, however, and the peace between Wilson and the paleoanthropologists did not last very long. Just as everyone was finally getting comfortable with the timing of the human–chimpanzee split, Wilson turned his attention to the last major event in human evolution, the origin of modern humans, our species *Homo sapiens*, and he dropped another bomb.

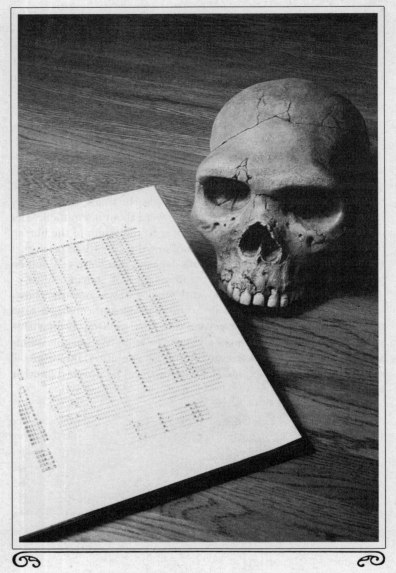

Figure 13.1 Neanderthal Forensics. A Neanderthal skull (a reconstruction) and a partial DNA sequence obtained from the original specimen. Composed by Jamie Carroll.

~ 13 ~
CSI: Neander Valley

> Every great advance in science has issued from
> a new audacity of imagination.
> — *John Dewey*, The Quest for Certainty *(1929)*

EIGHT MILES EAST OF Dusseldorf, Germany, the Düssel River cuts through deposits of limestone laid down more than 300 million years ago. Named in honor of Joachim Neander, a seventeenth-century poet, teacher, and composer of hymns, the Neander Valley wall was dotted with caves and rock shelters until the mid-1800s. The demands of the Prussian construction industry then led to quarrying operations on a large scale, and entire caves were removed.

One day in August 1856, the clay floor was being cleared from a small grotto called Feldhofer Cave to prevent its contaminating the high-grade limestone. As chunks of clay were tossed some 60 feet down to the valley floor, a number of bones and part of a skull were exposed. One of the owners of the quarry told the workers to watch out for more bones. Altogether, fifteen skeletal pieces and the skull were recovered. With a thick brow ridge and hefty, curved thighbone, the bones were first thought to be those of a cave bear, a common find in the region, so not much care was taken in their collection, and only the largest pieces were kept. Nevertheless, Carl Fuhlrott, a local schoolteacher and naturalist, was invited to visit the quarry. When he saw the bones, however, he identified them, not as those of a bear but of a human.

Fuhlrott noticed, though, that the bones had some unusual features and passed them on to Professor D. Schaafhausen in Bonn for a more expert appraisal. Schaafhausen agreed they were not typical human bones and concluded that they represented a natural form not previously seen, even in the "most barbarous races," and "that these remarkable human remains belonged to a period antecedent to the time of the Celts and Germans." Furthermore, Schaafhausen thought that "it was beyond doubt that these human relics were traceable to a period at which the latest animals of the diluvium still existed," by which he meant that this human coexisted with animals such as the cave bear, which had gone extinct at the time of some catastrophic flood.

Being 1857, just two years before Darwin's great opus appeared, the interpretations of fossil remains inspired many theories, and the Feldhofer bones even more so. Rudolf Virchow, then the leading German pathologist, disagreed with Schaafhausen and declared the bones those of a human deformed by rickets. The anatomist F. Meyer rejected both explanations and concluded that the bent leg bones and damaged elbow were just the decades-old remains of a Cossack cavalryman who had been injured in a battle with Napoleon's forces and who crawled into the cave and died.

Thomas Huxley took great interest in the "Neanderthal" man (*thal* meaning "valley" or "dale") and some considerable pleasure in poking holes in the skeptics' scenarios. Why, Huxley wondered, would a wounded soldier climb 60 feet up a cliff and remove his clothes and battle gear, and how would he then bury himself under two feet of clay *after he died?* No, Huxley concluded, the Neanderthal man was something different, an ancient human but one he thought was within the range of our species. William King, an Irish anatomist, went further than Huxley, and decided the Neanderthal man was related to but distinct from modern humans; it was a separate species, *Homo neanderthalensis.*

Subsequent discoveries in Belgium and France proved beyond a doubt that the skeleton from Feldhofer Cave was no deformed human or Cossack soldier but that of a distinct human form that was widely distributed across Europe, including as far south as Gibraltar, Spain, and Italy, as far west as England, and as far east as today's Iraq, Iran, and Uzbekistan. Their distinct features—a prominent brow ridge, large nasal cavity, and large heavy bodies—first inspired popular impressions of Neanderthals as brutes. Such inferior, apelike, cave-dwelling morons would therefore lie well "beneath" *Homo sapiens'* own scale of self-importance.

But over the past 150 years, the cartoonish, science fiction portrayals of Neanderthals has given way to a more objective study of Neanderthal history. We have many more, and more complete, Neanderthal specimens than of any other species beside ourselves, as well as considerable cultural materials from a wide range of sites.

Neanderthals occupy a critical chapter in the story of modern human origins, one that we know ended only about 28,000 years ago, but we are not so sure about some important details of the rest of the story. The central mystery (from our point of view, of course; it would be a different tale if they were still alive and a Neanderthal were writing this book) is how Neanderthals are related to us.

It is clear from the fossil record that we *Homo sapiens* and Neanderthals coexisted on the planet for a considerable length of time and cohabited Europe for perhaps 10,000 years before Neanderthals disappeared. But the questions of what transpired between us read more like the plot of the TV crime drama *CSI* than of highbrow science: What happened that led to their disappearance, and what role did *Homo sapiens* play? Did we murder and exterminate them? Or did *Homo sapiens* and Neanderthals meet in some French or Spanish cave for a little Ice Age romance? Is there a little Neanderthal in all or some of us?

Answers to these questions bear on the larger questions of just who we are, where we come from, and how we alone spread across the globe.

Ancestor or cousin?

In the first few decades of the 1900s, as the number of Neanderthal specimens increased and their study expanded, there was a shift in attitudes about our relationship to Neanderthals. As it became clear that these hominids had adapted to the generally colder climate of Europe as early as 300,000 years ago, long before the appearance of modern Europeans, one idea about their "disappearance" was simply that Neanderthals evolved into Europeans. This idea was first championed by Franz Weidenrich, a German paleontologist who worked for a period in China on Peking Man and other Chinese *Homo erectus* fossils. The Neanderthal-into-*sapiens* proposal was part of a larger theory that sought to reconcile the origin of modern humans and their diversity with the then-known distribution of *Homo erectus*.

Fossils of *H. erectus* were known from both Java (thanks to Eugène Dubois) and China. Weidenrich proposed that the Java form gave rise to native Australians and other similar populations in the region, the Chinese form to modern Chinese, and that *H. erectus* in Europe evolved via an *H. neanderthalensis* intermediate into modern Europeans. In this view, the racial differences among modern humans reflect ancient divisions in ancestral *H. erectus* populations.

With the discovery and dating of *H. erectus* in Africa by Louis Leakey, the time frame for this process of the parallel evolution of *H. sapiens* from *H. erectus* ancestors was pushed to at least a million years ago. This scenario is known as the "multiregional" model of modern human origins (Figure 13.2, left). Subsequent incarnations of this idea, advanced most notably by Milford Wolpoff, of the University of

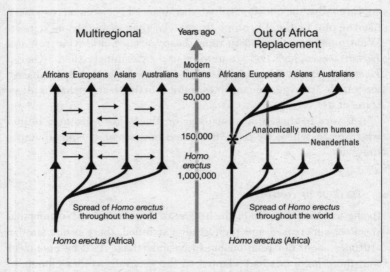

Figure 13.2 Two hypotheses of human origins. Left, the multiregional model posits that modern human populations are largely descended from archaic *Homo erectus* populations that were established long ago in different regions. In this model, Europeans would be the descendants of Neanderthals. Right, in the Out of Africa Replacement model, anatomically modern humans arose first in Africa 150,000–200,000 years ago and subsequently emigrated into different regions, replacing whichever humans were present with little or no interbreeding. Drawn by Leanne Olds.

Michigan, allow that some mating and thus gene flow occurred between different populations. But the distinguishing feature of this view is that it stresses continuity between the populations of ancient *H. erectus* and modern *H. sapiens* in each part of the Old World. This includes the continuity of some physical characters.

The difficulty some saw with the multiregional model, however, was in fact the lack of continuity between, for example, Neanderthals and modern Europeans, or Chinese *H. erectus* and modern Chinese people. Examination of early European *H. sapiens* such as the Cro-Magnons, which appeared about 40,000 years ago, did not reveal obvious anatomical connections to Neanderthals. Instead, the European fossil record suggested to some paleontologists that *Homo sapiens* simply *replaced* Neanderthals. If so, Neanderthals were not our ancestors, but a cousin of some sort, with whom we shared distant grandparents, and therefore a species that had been evolving separately from us for a considerable time. One intriguing proposal, made in light of Neanderthal's distribution, was that an ancestral human population split into two several hundred thousand years ago, separated by the expanding Sahara, with the northern population evolving into Neanderthals, and the southern population into *H. sapiens*.

Just like the earlier fossil record of apes and hominids, the weight of evidence hinged on interpretations and reinterpretations of morphological features of existing and new fossil finds. That was until 1987, when Allan Wilson got into the act.

Allan and Eve

The pounding that Wilson and Sarich took over their molecular clock–based estimates of the chimp-human split did nothing to discourage either man's interest in the questions of human evolution. Wilson spent part of a sabbatical in Kenya to learn more about the fossil evidence, and he had a deep interest in the evolution of the human brain and behavior and their link to origins of language. Linguists had identified some shared characteristics of modern languages, which Wilson took to indicate a shared, and recent, origin of all modern humans. Still dissatisfied with a strictly morphological approach to human history, he sought new ways of tackling the question.

For several years, Wilson had been exploiting the properties of mitochondrial DNA to investigate species relationships and histories.

The mitochondria in our cells, and those of other organisms with cell nuclei (called eukaryotes), harbor a small chromosome that encodes thirty-seven genes involved in mitochondrial function. Several properties of mitochondrial DNA (mtDNA) make it useful for evolutionary studies. First, it is abundant and can be purified and analyzed separately from nuclear DNA. Second, it mutates at a faster rate than nuclear DNA, so for any given time period, especially short periods (on the order of tens to hundreds of thousands of years), there are more ticks of the mitochondrial clock with which to measure past events. And third, it is generally inherited only from one's mother, and it can track pedigrees better than nuclear genes, which come from both parents and get shuffled by recombination.

Wilson decided to see what mtDNA could reveal about the timing and pattern of modern human diversification. With Rebecca Cann and Mark Stoneking, they examined mtDNA from 147 people representing five geographic regions, including Asians, Africans (mostly African Americans), Caucasians (from Europe, North Africa, and the Middle East), aboriginal Australians, and aboriginal New Guineans. They looked for various sequences in the mtDNA, and from similarities and differences between populations, they constructed evolutionary trees to decipher the most likely genealogy of the people they sampled. They reported that their trees had two main branches: one exclusively of Africans, and one of all five groups including Africans. From this pattern, they deduced that modern human mtDNA arose in and then spread out from Africa.

They then sought to establish a time scale for the origin of modern humans. They knew from previous work that the mean rate of divergence of mtDNA in most animals was about 2–4 percent per million years. In their samples, the average level of divergence of all mtDNA from a common ancestral type was about 0.57 percent, which implied that the common ancestor of all surviving mtDNA types lived just 140,000 to 290,000 years ago.

As Cann, Stoneking, and Wilson put it: "The tree . . . and the associated time scale . . . fits with one view of the fossil record: that transformation of archaic to anatomically modern forms of *Homo sapiens* occurred first in Africa, about 100,000 to 140,000 years ago, and that all present-day humans are descendants of that African population."

They underscored that these data were not consistent with the multiregional hypothesis; that would have predicted much greater genetic differences between human populations than Wilson's team measured (the older their separate ancestors, the greater the number of DNA changes that would have piled up over time). Furthermore, if there was some mixing between resident archaic populations in different parts of the world and newly arriving *Homo sapiens* from Africa, they would have expected to find more types of mtDNA in Asians than in Africans. Instead, they found the opposite. The African populations were the most different from one another, indicating that they had been evolving the longest and were the oldest modern human population. They proposed that "*Homo erectus* in Asia was replaced without much mixing with the invading *Homo sapiens* from Africa."

While they would later write in a scientific paper that their proposal "attracted much attention," "the shit hit the fan" would be a more apt description. This time, not only did molecular data again arouse the animosity of some paleontologists, but the discovery and debate quickly became very public.

Because of the maternal inheritance of human mtDNA, there must have been, logically speaking, an ultimate maternal ancestor, the mother of us all. What Wilson, Cann, and Stoneking affectionately dubbed "the Lucky Mother," whose mtDNA lineage survived and gave rise to all modern mtDNA types, the press quickly dubbed "Eve." The biblical allusion aroused both great interest, placing the mtDNA story on the cover of *Newsweek* magazine, and great ire in religious circles, not to mention among those folks who were not too keen to hear about their recent African ancestry.

That is not to say there were no legitimate scientific concerns; there were. The statistical methods and mathematical models used to reconstruct the history of molecules and populations were new and are still developing (and continue to be refined today). Several scientists questioned whether there was sufficient sampling of human populations and sufficient DNA sequence data or whether the best statistical methods were used in reaching the conclusions that Wilson's team did.

To proponents of the multiregional hypothesis, such questions were welcome. Wolpoff and others were vehemently opposed to what also became known as the "Out of Africa" Replacement theory (Figure 13.2, right). He organized a meeting of like-minded colleagues to "nail this 'Mitochondrial Eve' nonsense." He saw replacement as meaning just one thing—that *Homo sapiens* succeeded by violence: "It amounts to killer Africans with Rambo-like technology sweeping across the world and obliterating everybody they meet." Wilson was all too familiar with the view of some paleontologists that "fossils are the real evidence."

Could Wilson's methods or study have been faulty? The only way to know was to keep gathering data, and that is just what Wilson's team, and subsequently many others, did. An expanded analysis of more people (including more native Africans) and more mtDNA sequences, using different statistical tests, reached similar conclusions. The level of genetic diversity was highest in Africans, indicating again that they were the oldest *H. sapiens* population. Using several methods, the common human mtDNA ancestor was placed in the neighborhood of roughly 200,000 years ago.

Wilson's view was not at odds with those of all paleontologists. Christopher Stringer of the Museum of Natural History in London favored a single African origin of modern humans based on his interpretation of the fossil evidence. He was an early supporter of Wilson's genetic studies and cautioned colleagues that "paleoanthropologists who ignore the increasing wealth of genetic data on human population relationships will do so at their peril."

So to some degree history was repeating itself, with different camps polarized over which was the more reliable source of data—fossils or genes? What could help break the impasse, more fossils or more genes? Or could it be something else, such as genes from fossils themselves.

Bringing dead genes back to life

Wilson was adept at using the DNA of living species to peer back into the past. He often quipped to paleontologists that his approach held a crucial advantage because "living genes must have ancestors, whereas dead fossils may not have descendants." In other words, it was hard for paleontologists to be sure that they weren't looking at a dead end, a dead twig in the ever-branching evolutionary tree.

But for several years before the Eve story unfolded, Wilson and his colleagues were exploring an intriguing possibility that could bridge the two types of analysis—getting DNA directly from fossils.

To test the possibility, he decided to try working on preserved museum specimens. His first notable success was with a small chunk of tissue from the quagga, a relative of zebras that went extinct in 1883 (Figure 13.3). From a 140-year-old specimen in the Museum of Natural History at Mainz, West Germany, Wilson's team was able to extract enough mitochondrial DNA to obtain sequences from two genes. These sequences were found to be most similar to that of a zebra and indicated a time for the quagga-zebra split of a few million years ago. More important than this snippet of quagga genealogy, these were the first DNA sequences obtained from an extinct species (another "first" for Wilson). This step opened up the possibility that several fields "including paleontology, evolutionary biology, archeology, and forensic science may benefit."

Unbeknown to Wilson, in Uppsala, Sweden, a Ph.D. student was having similar thoughts. Ever since a trip to Egypt at the age of thirteen, Svante Pääbo had been drawn to ancient history. He enrolled at Uppsala to get a degree in Egyptology but soon discovered that the academic life was not all pyramids and mummies, for he spent most of his time deciphering Egyptian grammar. He decided to switch to medical school, then interrupted that path to study molecular immunology. As he started to learn the tools of cloning DNA, he realized that the techniques might also work on mummies. With the help of his former Egyptology professor and by working at night to keep his project secret, he extracted DNA from twenty-three mummies. He was even able to isolate a short segment of DNA from a 2,400-year-old mummy.

Just as he was writing up his success for publication, the Wilson team's quagga paper appeared in the journal *Nature*. Pääbo was disappointed that he had been scooped on cloning ancient DNA, but he greatly admired Wilson and sent him an advance copy of his mummy paper. Wilson was so impressed, he wrote back to Pääbo asking if he could spend a sabbatical in Pääbo's laboratory. Pääbo had to explain that not only did he not have a laboratory, he had not yet even finished his Ph.D. He turned the tables and asked if he could work with Wilson instead.

In 1987, the year the Eve story broke, Pääbo joined Wilson's lab; it was a fertile ground for new ideas and new technologies populated by many highly talented young scientists. Wilson's was the first ac-

Figure 13.3 The Quagga. The quagga is an extinct relative of the zebra. The first DNA sequences obtained from an extinct creature were taken from a museum specimen by Allan Wilson and his colleagues.

ademic lab to pursue a new technique called the polymerase chain reaction (PCR). In this procedure, segments of DNA sequence including entire genes were "amplified" from complex mixtures of DNA using a special enzyme that was stable at a high temperature. The new procedure allowed researchers to obtain large amounts of any specific piece of DNA and to quickly analyze the evolutionary changes that may have taken place in it between individuals or species. PCR revolutionized molecular evolutionary biology; it was the ideal tool for obtaining ancient DNA from ancient tissues, in which DNA was usually present in minuscule quantities at best, as well as being damaged and degraded. Using PCR, Pääbo was able to amplify mtDNA from both a 7,000-year-old human brain and the extinct marsupial wolf (also known as the thylacine). Pääbo and Wilson also collaborated on understanding the relationships of the New

Zealand kiwi to other flightless birds, including the extinct moa. But, sadly, Wilson died of leukemia before the study was completed.

The torch of ancient DNA studies was passed to Pääbo. Wilson had left his mark on his young protégé and instilled a particularly strong desire to study human evolution. In his own laboratory in Munich, Pääbo followed his mentor's path.

Secrets of the dead

He was asking the unthinkable. Pääbo pleaded with the curators of the Rhineland Museum in Bonn for samples. But these were not mummies or moas; Pääbo wanted Neanderthal bone, and not just any Neanderthal's bone but the bones of *the* Neanderthal: numero uno— the type specimen number 1—that had been found by those quarrymen in 1856. And he didn't want to just see the bone, he wanted a piece cut out so he could grind it up and extract DNA. That's right, he wanted to pulverize a slice of a national treasure. The curators were understandably reluctant. But Pääbo persisted, and in 1996 he got his slice—a half-inch chunk carved out of the upper right arm bone (Figure 13.4).

If Neanderthals contributed to Europeans or to any other *H. sapiens*—if Pääbo's expert team could extract and sequence 42,000-year-old DNA—this ought to settle it. Bona fide sequences had not been obtained from anything that old except from some mammoths that had been well preserved in permafrost. Because the Neanderthal bones were handled without protection by their discoverers, museum curators, and who knows how many more people, the contamination of ancient samples with modern human DNA was a terrible problem. PCR was great at picking up small traces of DNA, sometimes too good. The literature was filled with bogus results from contamination. Pääbo could not afford to be fooled. The stakes were very high: this was the most ambitious and critical test for ancient DNA. If they failed, there might not be a second chance, for the fossils were too precious to keep slicing them to pieces.

Pääbo had learned over the years just how prevalent the modern human contamination of ancient samples was and devised every precaution to minimize it. Protective clothing had to be worn, instruments were treated with hydrochloric acid and rinsed with sterile water, the samples (removed with a drill saw) were placed into a sterile tube, and all of the DNA work was carried out in a lab specially designed for handling archaeological specimens and where UV irradiation and other measures were used to avoid contamination.

The bone pieces were ground into a powder, and the tiny amount of DNA present was extracted with a series of solvents. Then, the PCR technique was used to amplify short segments of the mtDNA. Next, the sequences of bases in the amplified DNA was determined. The moment of truth arrived and was for Pääbo "one of these really cool moments of life." He saw immediately that the Neanderthal sequences were very different from those of modern humans.

But they had to be sure. Pääbo wanted the whole procedure repeated, not just many times in his lab, but in another lab altogether. He sent a sample to Mark Stoneking, a Wilson lab teammate and coauthor of the Out of Africa papers, then at the Anthropological Genetics Laboratory at Penn State. Stoneking obtained the same sequences as Pääbo's team.

The results showed that there was no evidence that Neanderthals had contributed mtDNA to modern humans, which would be expected by the multiregional sequence. Furthermore, the Neanderthal sequence was not any more related to that of modern Europeans than to that of any other human population, which indicated that the Neanderthal line split off from humans well before different human populations diverged. The Neanderthal sequence differed on average at 27 of 378 positions from those of modern humans, while different human populations (e.g., African, European) differ from each other on average at about 8 positions. Using a molecular clock to calculate the Neanderthal–*H. sapiens* split, Pääbo's team arrived at an estimate of 550,000 to 690,000 years ago for the divergence of human and Neanderthal mtDNA.

A commentary accompanying the publication of these findings hailed the work as "a landmark discovery . . . arguably the greatest achievement so far in the field of ancient DNA research." For paleoanthropology, it marked another turning point in terms of both practice and theory. Pääbo's team showed that unique genetic information could be extracted from human fossils and that scientists had a whole new tool beyond comparative anatomy and radiometric dating for deciphering human origins. And it was another blow to the multiregional hypothesis and further support for the Out of Africa Replacement model.

Figure 13.4 **Getting DNA from a Neanderthal.** A slice was taken from the upper arm bone of the original Neanderthal-type specimen from Feldhofer Cave and processed for DNA extraction and sequencing. Reprinted from M. Krings et al. (1997), *Cell* 90: 19–30, with permission from Elsevier.

That is, if it was correct. After all, the Feldhofer bones were just one sample, and the possibility remained that it may not have been representative of all Neanderthals. It was therefore crucial to get more DNA sequences and more specimens. Pääbo's team did both: they extended the amount of DNA sequence from the Feldhofer-type specimen, and they and others obtained DNA sequences from Neanderthal specimens from Croatia, Russia, Belgium, and France. All of the Neanderthal remains yielded mtDNA sequences that were very similar to one another and were not represented in human mtDNA.

But do these findings rule out any interbreeding of Neanderthals and *Homo sapiens* or any contribution of Neanderthals to the modern human gene pool? They do not, and the reasons come from a bit more consideration of what mtDNA can tell us. MtDNA is maternally inherited, so it can reveal only possible breeding between Neanderthal females and modern human males. Furthermore, a given mtDNA lineage can go extinct without the whole species going extinct. So it is possible that the few Neanderthals sampled had no modern human descendants, but other Neanderthals might have.

Another way to look for possible mixing of Neanderthal and *Homo sapiens,* and perhaps to minimize the time available for extinction of an mtDNA lineage to occur, is to look at the DNA of early modern humans in Europe that lived closer in time to Neanderthals. Pääbo's team examined DNA from five early modern human specimens from the Czech Republic and France, including a Cro-Magnon sample from about 23,000–25,000 years ago. There were no diagnostic Neanderthal mtDNA sequences in these early *Homo sapiens* specimens.

While this is yet more supporting evidence that Neanderthals did not mix with modern humans or contribute to modern human evolution, the inheritance of mtDNA is just one part of the picture. The genes that govern most of our body processes and our anatomy reside in nuclear DNA. But because this DNA exists in only two copies per cell, as opposed to the 100–10,000 copies of mtDNA per cell, and because ancient DNA is degraded into very short pieces, the hope for obtaining much, if any, nuclear DNA sequences from Neanderthals seemed dim at best. In fact, when Pääbo's first successes with mtDNA were hailed, the view at the time was that "it is not within our power . . . to retrieve the great majority of lost genetic information."

We are powerless no more. Dramatic advances in DNA sequencing technology, driven by the demand in medical genetics for ever faster and cheaper methods, have provided the added dividend of not just the contemplation of the sequencing of some Neanderthal nuclear DNA but of the *entire genome*—the entire 3 billion DNA base pairs of an extinct human. Two teams, one led by Svante Pääbo (now at the Max Planck Institute for Evolutionary Anthropology in Leipzig, Germany), the second by Eddy Rubin at the U.S. Department of Energy Joint Genome Institute in Walnut Creek, California, have embarked on this new adventure. The technical obstacles are formidable—degradation and chemical decomposition of the DNA leave only short segments of about fifty letters of DNA text, and many tens of millions of these short pieces will need to be sequenced and assembled in their correct order, with contamination by modern human DNA a constant concern.

Nevertheless, the first nuclear DNA sequences were obtained in 2006, and they tell a story consistent with the previous mtDNA analysis. From roughly 65,000 base pairs of sequence—just a minuscule fraction of the 3 billion base pairs to be determined but a hundredfold increase in the amount of data relative to the previous decade's output of mtDNA studies—the Neanderthal–*Homo sapiens* split was again pegged at several hundred thousand years ago. Moreover, no evidence of Neanderthal contribution to the modern human gene pool was detected. If Neanderthals and *Homo sapiens* did commingle, their great-great-great-great-great- etc.-grandchildren aren't revealing that family secret.

A very cold case

If we are not their descendants, just what is our history, and what happened to the Neanderthals?

Based upon the fossil, DNA, and cultural records, a picture is emerging with respect to the general chronology of our origins and Neanderthal history (Figure 13.5). It now appears that the most recent common ancestor of Neanderthals and *Homo sapiens* arose as early as 700,000 years ago, and that the two populations split 300,000–400,000 years ago, well before the origin of the earliest known modern *Homo sapiens* (a fossil from Herto, Ethiopia, dated at about 160,000 years old). For perhaps 300,000 years or more, anatomically distinct Neanderthals populated Europe and reached western Asia 150,000 years ago. Their known geographic range was recently extended more than 1,000 miles eastward, to the Altai mountains of southern Siberia, in yet another coup for Neanderthal DNA forensics. The identity of some fragmentary fossil remains from a cave was disputed and uncertain until DNA sequences were obtained by Pääbo and his colleagues that indicated they belonged to a Neanderthal.

Sometime around 60,000 years ago, anatomically modern humans began a series of migrations out of Africa into Asia and Australia and reached parts of Europe about 40,000 years ago. Whereas only 50,000 years ago the only humans living across a wide belt of Europe and Asia were Neanderthals, by 30,000 years ago most humans across this region were *Homo sapiens*, with the last remaining Neanderthals occupying pockets of far southern Europe until they vanished altogether by about 28,000 years ago (Figure 13.5).

Many scenarios have been put forward to explain the disappearance of the Neanderthals. The most sensational, of course, is that our ancestors violently wiped them out. Other ideas include the possibility that arriving *Homo sapiens* brought diseases to which Neanderthals were susceptible, as has occurred many times when different human populations come into contact. Other theories suggest that Neanderthals were in some ways inferior to the arriving *Homo sapiens* in their degree of technological sophistication, or communication skills, or degree of social organization, and were then outcompeted for available resources.

While the mystery surrounding the death of the Neanderthals is certain to inspire speculation and debate for a long time to come, there appear to be some clues in the climatic, biological, and cultural records of this critical period 50,000 to 30,000 years ago.

The Pleistocene epoch (1.8 million to 11.5 thousand years ago) was marked by repeated glaciations that had dramatic effects on vegetation, wildlife, and the landscape. Neanderthals as a species had endured long periods of harsh and variable climate during the Middle (781–126 thousand years ago) and Late (126–11.5 thousand years

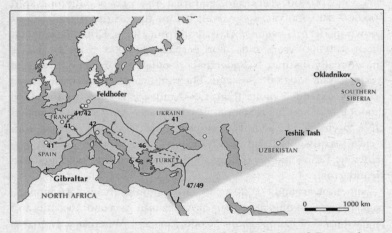

Figure 13.5. **Neanderthal's range and the invasion of Europe by** *Homo sapiens.* The range of Neanderthals based on fossil and DNA evidence is shaded in gray. Individual sites are shown as open circles and the locations of some key sites are indicated, including the original Feldhofer site, and Gibraltar, where the youngest remains have been found. The inferred migration routes of *Homo sapiens* from Africa through the Middle East and into Europe are shown as arrows and dotted arrows and the dates of arrival indicated by numbers in bold. The figure is based on information in J. Krause et al. (2007), *Nature* 449: 901–4, and P. Mellars (2006), *Nature* 439: 931–35. Figure drawn by Leanne Olds.

ago) Pleistocene, and their bodies reflected their adaptation to a colder climate. They were certainly good big-game hunters, and Neanderthal fossils reflect their rugged lifestyle in another way—they often show multiple healed fractures, so many, in fact, that their skeletons have drawn analogies to those of modern rodeo riders.

However, the paleoclimatic record indicates that Europe experienced dramatic swings in climate during the period 50,000–30,000 years ago. Rapid oscillations in climate, occurring on the scale of centuries, caused swift expansions and contractions of various kinds of habitat. Cold conditions caused the reduction of woodlands and the formation of open plains and tundra. The fossil record over this period documents shifting distributions and numbers of mammals, including the disappearance of some animals of warmer times, such as the straight-tusked elephant. Neanderthals clearly had to adapt to a changing dinner menu.

As the range of treeless vegetation spread, the available game changed, and with it the most effective hunting techniques also had to change. Neanderthals operated out of woodlands and used forest cover to stalk and ambush grazing mammals at close range. But as the plains opened up, moving herds of bison, red deer, and horses were better stalked by humans with long-range mobility, who were able to live away from woods and caves and were equipped with hunting tools better suited for long-range attack, such as pointed blades and spears—in short, the newly arriving *Homo sapiens*.

In the cultural record, the transformation in technology over this time period is dramatic. By about 41,000 years ago, close to the time that individual died in the Feldhofer Cave in the Neander Valley, the "Aurignacian" culture (named for an archaeological site in France) arrived in central and western Europe. The artifacts of this invasion include the first complex and carefully shaped ivory, antler, and bone tools; carefully carved stone and ivory beads and personal ornaments; marine shells that had been carried from a long distance; perforated animal teeth; ivory statuettes of human and animal figures; elaborate cave art; and, perhaps most important of all, carefully shaped bladelets for the tips and barbs of spears and arrows. All of these objects mark a sharp break from preexisting Neanderthal technologies —and some evidence of the new technologies can be traced back to at least 70,000 years ago in African populations.

As temperatures became colder, innovation and social organization were essential to the survival of human populations. The use of plant fibers to make fishing nets and snares for small mammals and birds expanded the diet of *Homo sapiens*, while the development of trade networks afforded the mobility to follow migrating game through different territories and seasons. By 30,000 years ago, large populations of *Homo sapiens* occupied individual sites. Ecological changes and competition, then, with or without direct conflict, appear to have pushed Neanderthals to the fringes of their former range and resulted in their ultimate replacement by *Homo sapiens*.

Why was our species and not the Neanderthals able to adapt? Were Neanderthals inferior in some biological ways—in their capacity for communication, thought, or planning? Why did our ancestors, not Neanderthals, develop certain technologies and cultural traditions? In short, what made the differences between *us* and *them?* Biological explanations for our different fates seemed to be restricted to the realm of speculation, beyond the reach of hard physical evidence, just a year or two ago, but no longer. Comparisons of the sequence of the genome of our recently extinct cousin with ours may be able to tell us in what ways we are different and, just as important, how we are the same. A new window on Neanderthal biology and modern human origins is opening.

ᘒ AFTERWORD ᘒ
The Shape of Things to Come

Mysteries are not necessarily miracles.
—*Johann Wolfgang von Goethe (1749–1832)*

DESPITE ALL OF THE PRIVATIONS they endured, nearly all the adventurers we have followed expressed some sadness at returning to civilization and leaving behind the jungles, deserts, badlands, canyons, and mountains that had beckoned them. I, too, feel a bit forlorn at leaving behind these remarkable pioneers with whom I have spent so many enjoyable days. I hope that you might feel a touch the same at reaching the last of these stories. If so, then I have done my job.

At the end of any journey or on the occasion of an anniversary, it is natural to reflect on how far one has traveled and to ponder what lies ahead. So, as we leave the stories of these adventures, and mark several anniversaries of major works and discoveries, let's pause to reflect upon how far we have come and what may lie in the future. I don't mean this in terms of physical distances and places but in regard to the more profound impacts of these discoveries and ideas—on our perception of our place in the world. In this brief afterword, I will focus on two questions. First, where has the Darwinian revolution led us? And second, are things of a similar magnitude still to be discovered?

The world into which Darwin led us

Fifty years ago, on the centennial of the publication of *The Origin of Species*, George Gaylord Simpson, a paleontologist and co-architect of the "Modern Synthesis" of evolutionary theory, wrote a masterful essay for the journal *Science* entitled "The World into Which Darwin Led Us." He examined how the Darwinian revolution changed, completely and forever, long-cherished concepts of ourselves.

Most, if not all, of what Simpson wrote holds true today, only more so. His main points, and the artful way he phrased them, bear repeating. He explained that the Darwinian revolution extended a transformation in thought begun by early astronomers who "have finally located us on an insignificant mote in an incomprehensible vastness—surely a world awesomely different from that in which our ancestors lived not so many generations ago." Geologists then extended this vastness to the dimension of time by deducing that the age of the earth was in the millions of years (now known to be 4.5 billion years). This great age, Simpson observed, described "a world very different from one conceived as less than 6000 years old."

The Darwinian revolution then delivered three more blows to the perceptions of our place and purpose in nature.

First, it revealed that the world and the universe are hostile places, not the peaceful, orderly realms perceived by Darwin's predecessors, such as Humboldt.

Second, Darwin's new picture of ancestry meant that humans have no special status other than our definition as a distinct species of animal. Our kinship with amoeba, tapeworms, fleas, and monkeys represented to Simpson "togetherness and brotherhood with a vengeance, beyond the wildest dreams of copy writers or of theologians."

And third, the struggle for life made it extremely improbable that anything in the world exists specifically for our benefit or ill. Simpson offered: "It is no more true that fruits, for instance, evolved for the delectation of men than that men evolved for the delectation of tigers."

While the main elements of this philosophical revolution were so clear to Simpson fifty years ago, the scientific revolution that Darwin started did not end in 1959. Discoveries over the ensuing decades have even shaken Simpson's sober view of the world. The universe, for instance, is in fact more hostile and uncaring than he knew. While well aware of the prevalence of extinction, Simpson saw the geological and fossil record as documenting steady, gradual, orderly change. We now know that the face of the earth has been remodeled and the planet's inhabitants extinguished by cataclysmic events such as the

K-T asteroid impact (Chapter 8). Catastrophic scenarios were long disdained by geologists as unmodern and unscientific, until Chicxulub. While Simpson would have surely been startled by this revelation, its implications would reinforce his view that life does not evolve toward a goal. Regarding the contradiction between the notion of life evolving progressively and the pervasiveness of extinction, Simpson noted, "If that is a foreordained plan, it is an oddly ineffective one."

In the last fifty years, our pictures of both human origins—who we are and where we came from—and of the mechanisms of evolution have been transformed. The first of the Leakeys' great hominid fossil discoveries happened that very year of 1959 (Chapter 11). They triggered a revolution in hominid paleontology that took the study of our origins back to Africa (as Darwin had surmised, but because of Dubois' discovery had many focused on Asia).

The cracking of the genetic code shortly thereafter led, as we have seen here (Chapters 12 and 13), to entirely new ways of deciphering human history. DNA has delivered the crucial evidence that we are more recently diverged from a common ancestor with, and genetically even more closely related to, chimpanzees than most once thought, and many feared. DNA evidence has also conclusively revealed that we are all Africans (and not the descendants of Neanderthals).

The DNA revolution has also transformed our understanding of the evolution of molecules and organisms. The discovery of the molecular clock revealed that molecular change and physical change are, contrary to Simpson's intuition and objections, uncoupled. And advances in DNA analysis have enabled biologists to see the operation of natural selection at its most fundamental level.

It could be said then that the first great leaps toward solving "the mystery of mysteries" (the origin of species) were taken by Darwin and Wallace 150 years ago, while the greatest strides in answering "the question of questions for mankind" (our origins) have come in the last fifty years.

So what about the next fifty years? Using the last five decades as a guide, it would be folly to claim the revolution is complete. More fossils and more information unearthed from DNA will surprise and enrich us. But it is worth asking, now that we have a solid grasp of evolution and our origins, are there other open questions of a similar magnitude to those that have occupied the last 150 years?

I submit that the outstanding issue, and perhaps the greatest mystery of mysteries and question of questions, is the ultimate matter of origins—the origin of life in the universe and on Earth.

One in a billion trillion?

Are there other worlds that could carry life?

Is there, in fact, life elsewhere?

If so, what kind of life is it?

These are not new questions. For millennia humans have looked to the stars and wondered what lies "out there." Ever since Copernicus declared, in 1543, that our planet orbits the sun, astronomers have been adjusting our perceptions of our place in the universe.

In 1584, the Catholic monk Giordano Bruno asserted that there were "countless suns and countless earths all rotating around their suns." He was charged with heresy and burned at the stake in 1600.

(And here I thought peer review these days was unforgiving.)

Bruno's story does have a happy ending, however. Exactly 400 years later, in 2000, the Church officially expressed its "profound sorrow" for roasting Bruno in Rome's Piazza dei Fiori. It is hard to know for certain what delayed the apology; the Vatican is pretty tight-lipped about such things. Perhaps it was Bruno's assertion that the universe was infinite, but I am betting that his characterization of monks as "holy idiots" had something to do with it.

We can now say there are other suns and planets out there (I'll discuss the evidence shortly) without incurring the wrath of the holy idiots. Planets are not the chief issue; the presence of life is. We can chuckle at the quaint attitudes of 1600, but what would the world's reaction be to firm evidence of life elsewhere? Judging from the reaction to the astronomers' demotion of the earth as the center of the solar system and Darwin's demotion of humans as the species for whom the world was shaped, what impact would the demotion of the earth as the sole province of life have on our perceptions? One origin of life might be perceived as a miracle, but multiple origins? To scientists, that would be confirmation that life is a common product of planetary chemistry. Remarkable, yes, but not a miracle.

The wider impact of such a discovery is not lost on the scientific community. In 2001, the National Research Council, an advisory body of top scientists, stated: "The discovery of life on another planet is potentially one of the most important scientific advances of this century, let alone this decade, and it would have enormous philosophical implications." It would certainly be a complete break from the view of the world and the cosmos that prevailed just a century and a half ago.

The prospect of extraterrestrial life was contemplated by Humboldt but rejected in his *Kosmos* (1845):

> The starry vault and the wide expanse of the heavens belong to a picture of the universe, in which the magnitude of masses, the number of congregated suns and faintly glimmering nebulae, although they excite our wonder and astonishment, manifest themselves to us in apparent isolation, and as utterly devoid of all evidence of their being the scenes of organic life.

Darwin was cautious on the matter; he had his hands full with challenges to his evolutionary theory. In the second edition of *The Origin of Species*, he added the words "by the Creator" to the last line: "life . . . having been originally breathed by the Creator into a few forms or into one." He later expressed regret for bowing to pressure to put a Creator back into the scheme of life's origins. More revealingly, in one of the last letters he wrote before he died, he speculated that life "will hereafter be shown to be a part, or consequence of some general law."

The possibility of extraterrestrial life received its greatest boost in popular circles with the birth of the science fiction genre, led by Jules Verne (1828–1905) and H. G. Wells (1866–1946). The latter was a biology student of and inspired by Thomas Huxley. It was Huxley's view of a hostile universe shaping mankind that colored Wells' *Time Machine* (1895), while his *War of the Worlds* (1898) depicted an alien invasion of England from Mars.

So far, Wells' vision of Martian invaders has not been borne out, but what do we actually know about life elsewhere?

There was quite a stir in 1996 with the claim that a Martian meteorite found in the Antarctic (in 1984) bore evidence of microbial life. It would have been an amazing stroke of luck—to have Martian life come to us instead of our having to figure out how to find it. But further scrutiny has led most scientists to conclude that the minute shapes in the meteorite were caused by nonbiological processes.

And there lies the rub. How exactly should alien life look if we did find it? To put it in Huxleyesque terms: Should it be Earthlike in form? Or will it be unlike anything found on Earth?

The more systematic approach, rather than waiting for a lucky meteorite, is to go look for life directly. There have been several missions to Mars that have conducted increasingly sophisticated scientific surveys of the Martian surface. The general idea is to look for water and other geological features that might reflect the existence of life at some time in the planet's history. While there is good evidence that water has flowed at times, the prospect of life there at any time has been growing dimmer.

For the sake of discussion, let's assume that we've struck out with our nearest neighbor. What about planets elsewhere? Answering that question depends on the parameters one thinks are conducive to or essential for life. Over the years, a variety of ideas have been put forward. While there is not unanimous agreement (which is good, since we don't know what we are looking for), some consensus characteristics have emerged. In general, the thinking is that planets need to be of sufficient size to maintain an atmosphere, to be rocky (not a gas giant), to have or have had active geology (volcanoes, etc.), to bear liquid water, and to be close enough to a sun to be warm but not so close as to be extremely hot or bombarded with too much radiation.

What are the odds of finding these conditions? The astronomers have been wrestling with that question. Aided by the Hubble telescope and many other technological developments, they have been peering ever more deeply into the universe. The first planet orbiting a star similar to the sun was just discovered in 1995. Since then, a couple of hundred "new" planets have been reported. Each of these discoveries required long periods of sustained observation, and they represent an infinitesimal sample of the universe.

To figure out how common Earthlike planets may be, we have to crunch some truly astronomical numbers. Let's start with the number of galaxies in the universe; a conservative estimate is 100 billion, that's 100,000,000,000 and is written in shorthand as 10 to the eleventh power (10^{11}). Now, how many stars are there in a galaxy? A good stab would put that figure at about 100 billion (10^{11}) as well. That means that the total number of stars would be 10^{11} times 10^{11}, or 10^{22} stars. Not all stars have planets around them, but NASA reports that about 7 percent of nearby stars harbor a giant planet, and that the number of planets around each star increases as the planet's mass decreases toward the size of Earth. From these considerations, the current estimate is that there are 10^{21} Earthlike planets, 10^{10} in our galaxy alone. That's 1 billion trillion Earthlike planets in the universe, and 10 billion such planets in our galaxy. So what are the odds that life evolved only here on earth? Are we one in a billion trillion?

You are free to take your own stab at that question.

But I will divulge that every scientist I have asked who is knowledgeable in this arena thinks life elsewhere is very, very likely—a virtual certainty. So the question of extraterrestrial life in many minds is not so much a matter of "if" but of "what is it like?"

Are there dinosaurs, ape-men, and DNA-based life out there?

The shape of things to come

Chris McKay, a planetary scientist at NASA, while acknowledging that there is no solid evidence yet of life elsewhere, thinks "several factors suggest it is common. Organic material [carbon-based molecules] is widespread in the interstellar medium and in our solar system." He notes, "On Earth, microbial life appeared very quickly— probably before 3.8 billion years ago" (that estimate varies from around 3.5 billion–3.8 billion years ago). Coupled with the abundance of planets and the ability of microbes to survive in a wide range of habitats, McKay concludes that "microbial life . . . is widespread in the stellar neighborhood."

Peter Ward, a geologist at the University of Washington and coauthor of the book *Rare Earth*, agrees that microbes are common in the universe but argues that complex life, multicellular forms like plants and animals, are far more rare. Ward and others give a lot of weight to the fact that it took 3 billion more years of evolution and major changes in the oceans and atmosphere before larger, complex forms appeared. In this view, finches, butterflies, trilobites, and redwoods are not a given.

And if they are not inevitable, what of intelligent life? Or hominoids? Or book authors? Draw your own conclusions.

And what about the chemical bases of life? Must life rely on a DNA-based system? Some scientists are not convinced that our carbon-in-water system is the only option nor that DNA is necessary. Frank Drake, a longtime proponent of extraterrestrial exploration, thinks that "we are over our heads predicting what [chemistry] might have taken place on other worlds."

Whatever its shape or chemistry, there is one important expectation that many if not all biologists share, as Simpson also did, about life in the universe. It is that wherever life has arisen, it has evolved by the two principles Darwin formulated—by descent with modification and natural selection. These principles concerning the continuity of forms, and competition among replicating forms, are expected to be truly universal.

Andy Knoll, a Harvard paleontologist, the author of *Life on a Young Planet*, and a Mars Rover scientist, cautioned me in a recent exchange that "talk is cheap, exploration and discovery is hard." So what are the imminent prospects for exploration and discovery? What new adventures lie ahead?

In addition to many ongoing efforts, over the next fifteen years NASA will launch several new missions to search for and to characterize new worlds. Its Terrestrial Planet Finder observatories, for example, will have a hundred times the imaging power of the Hubble telescope and will enable the search for planets with atmospheres containing carbon dioxide, water, and ozone. The European Space Agency has on its drawing boards a mission beginning in 2015 to search for Earthlike planets and evidence for life on such bodies. The current design of three telescopes flying in formation is the centerpiece of their program that has been christened "The Darwin Mission."

Whether in the next decade or in the next century, and no matter what its shape or chemistry, when no longer the realm of speculation and science fiction, the conclusive discovery of extraterrestrial life will be another profound turning point in our perception of our place in the universe. Even such a marvelous achievement, however, will not be the climax of our exploration but the opening of a new era, for as Wells foretold in his 1933 novel, *The Shape of Things to Come*:

for man, no rest and no ending. He must go on, conquest beyond conquest. First this little planet and all its winds and ways, and then all of the laws of mind and matter that restrain him. Then the planets about him, and, at last, across immensities to the stars. And when he has conquered all the deep space, and all the mysteries of time, still he will be beginning.

✃ Acknowledgments ✃

This book would never have unfolded without the support, encouragement, patience, and keen critical input of my wife, Jamie Carroll. Jamie read every chapter from its earliest draft onward and provided great guidance on making the stories both interesting and readable. My sons, Patrick and Will, were also willing listeners as I tried out a few yarns on them.

I was lucky to be brought up in a storytelling family. Maybe it is an Irish trait, as we heard and told a lot of stories growing up (not all of them true). We took a lot longer than our neighbors to finish dinner, something my mother is very proud of to this day. So thanks Mom and Dad, Jim, Nan, and Pete for your stories, and the occasional chance to tell mine.

I am indebted to many other storytellers. First, there are the principals who not only bravely undertook these journeys but kept detailed records and often wrote about their experiences in the first person. And second, I would not have known about the great drama of Darwin's life and voyage, or Wallace's travails, or Dubois' determination, or the qualities and experiences of many of the other figures in this book were it not for the painstaking work and talents of many biographers, historians, and writers. I was educated and inspired by their storytelling, and I hope that my brief accounts do their subjects justice.

I am also very grateful to my friend Neil Shubin for the generous time he spent telling me the story of his adventures, for helping me to get it right, for the photos from his expedition, and for showing me *Tiktaalik*.

I am also indebted to several colleagues at the University of Wisconsin-Madison for their generous creative, critical, and logistical contributions. Leanne Olds created most of the original artwork for the book; Steve Paddock provided detailed comments and suggestions on the entire manuscript; and Megan McGlone secured the permissions for the images obtained from other sources. I was also particularly fortunate to have liberal access to the holdings of the University of Wisconsin library system. The researching of this book led me to many old and rare volumes; many thanks to Elsa Althen for providing access to the rare books collection of the Biology library.

I also thank the many museums and individuals who provided illustrations for the book.

Finally, I thank my agent, Russ Galen, for his encouragement and patience and for navigating the complexities presented by my writing some of this material for a student text; and my editor at Houghton Mifflin Harcourt, Andrea Schulz, for her enthusiastic support and critical input.

∾ Sources and Further Reading ∾

I have used a wide range of sources in researching the stories in this book. I decided not to use footnotes in order to avoid the cluttering and distraction that can entail. The source of each quote is identified here by page number. For each chapter, the books, articles, and websites cited are listed.

One bit of warning, should you wish to examine some of these stories in greater detail. In wading through the voluminous documents pertaining to certain chapters, I often found considerable redundancy, even in those materials written by the principal figures. For example, Wallace's dispatches from the field that were published as articles in journals are often lifted *in toto* in his *Malay Archipelago* and then again in his autobiography, *My Life*. Roy Chapman Andrews' accounts are also repeated many times in a succession of his books and in official accounts of his expedition. I needed to read all of the material to know what information was available, but if you look in several sources for any one story, be aware that extensive redundancy is possible. Looking at it another way, I read all of this stuff so you don't have to. You are most welcome.

I have provided some guidance by starring certain references as recommended reading.

1. Introduction: Humboldt's Gifts

BOOKS

Ceram, C. W. (1986). *Gods, Graves, and Scholars.* 2nd revised edition. New York: Random House.

Furneaux, R. (1969). *The Amazon: The Story of a Great River.* London: Hamish Hamilton.

McCullough, D. (1992). *Brave Companions: Portraits in History.* New York: Simon and Schuster.

Von Humboldt, A. (1850). *Aspects of Nature in Different Lands and Different Climates; with Scientific Elucidations.* Philadelphia: Lea and Blanchard.

ARTICLES

Bowler, P. J. (2002). Climb Chimborazo and See the World. *Science* 298: 63–64.

Browne, C. A. (1944). Alexander von Humboldt as historian of science in Latin America. *Isis* 35: 134–39.

Bunkse, E. V. (1981). Humboldt and an Aesthetic Tradition in Geography. *American Geographical Society of New York* 71: 129–45.

Coonen, L. P., and C. M. Porter (1976). Thomas Jefferson and American Biology. *Bioscience* 26: 745–50.

Crone, G. R. (1961). Alexander von Humboldt: Centenary Studies. *Geographical Journal* 127: 226–27.

De Terra, H. (1960). Alexander von Humboldt's Correspondence with Jefferson, Madison, and Gallatin. *Proceedings of the American Philosophical Society* 103: 783–806.

———. (1960). Motives and Consequences of Alexander von Humboldt's Visit to the United States. *Proceedings of the American Philosophical Society* 104: 314–16.

Dettelbach, M. (2001). Alexander Von Humboldt between enlightenment and romanticism. *Northeast Naturalist* Special Issue 1: 9–20.

Kettenmann, H. (1997). Alexander von Humboldt and the concept of animal electricity. *Trends in Neuroscience* 20: 239–42.

Knobloch, E. (2007). Alexander von Humboldt—The Explorer and the Scientist. *Entaurus* 49: 3–14.

Rowland, S. M. (2005). Thomas Jefferson, Megalonyx, and the status of paleontological thought in America at the close of the eighteenth century. *Geological Society of America Abstracts with Programs* 37: 406.

Schwarz, I. (2001). Alexander von Humboldt's visit to Washington and Philadelphia, his friendship with Jefferson, and his fascination with the United States. *Northeast Naturalist,* Special Issue 1: 43–56.

Von Hofsten, N. (1936). Ideas of Creation and Spontaneous Generation Prior to Darwin. *Isis* 25: 80–94.

Walls, L. D. (2001). Hero of Knowledge, Be Our Tribute Thine: Alexander von Humboldt in Victorian America. *Northeast Naturalist,* Special Issue 1: 121–34.

WEBSITES

Aber, J. S. (2003). History of Geology: Baron Friedrich W.K.H. Alexander von Humboldt. http://academic.emporia.edu/aberjame/histgeol/humboldt/humboldt.htm, accessed December 17, 2006.

"Fossils." *The Lewis and Clark Journey of Discovery.* http://www.nps.gov/archive/jeff/LewisClark2/TheJourney/Fossils.htm, accessed December 22, 2006.

Jewett, T. O. (2000). "Thomas Jefferson, Paleontologist." *Early America Review* 3.2: http://www.earlyamerica.com/review/2000_fall/jefferson_paleon.html, accessed December 22, 2006.

Lepenies, W. (May 31, 1999). "Alexander von Humboldt—His Past and His Present." Talk given at *Haus der Kulturen der Welt,* Berlin. *BerliNews: Wissenschaftsgeschichte,* accessed December 24, 2006.

Murphy, D. C. "Discovering the Great Claw: Part 1—The Giant Cat." *National Academy of Natural Sciences* http://www.ansp.org/museum/jefferson/megalonyx/history-01.php

"An Abstract of Mr. Emerson's remarks made at the celebration of the centennial anniversary of the birth of Alexander Von Humboldt." *The Complete Works of Ralph Waldo Emerson* http:www.rwe.org/comm./index2.php?option+com_content&task=view&id=94&itemid=298, accessed December 28, 2007.

SOURCES OF QUOTES

1   *"one of those wonders of the world":* Humboldt XXIV, *The Complete Works of Ralph Waldo Emerson.*
    "What trees! Coconut trees, fifty to sixty foot high": in McCullough (1992), 6–7.

2   *"I do not remember ever having received a more dreadful shock":* in Furneaux (1969), 103.
    "I often tried, both insulated and uninsulated": ibid.
    "vomited up rubber balls for several hours": ibid., 102.
    "its taste of an agreeable bitter": ibid.

3   *"I felt very much stirred up":* in Schwarz (2001), 43.

4   *"I would love to talk to you about a subject":* de Terra (1960), 288.

5   *"Such is the economy of nature":* Murphy, Academy of Natural Sciences website.
    "In the present interior of our continent": ibid.
    "all the animals of the country generally": in Coonen and Porter (1976), 748.

7   *"the master of all branches of science":* Furneaux (1969), 97.
    "There are few heroes who lose so little": in McCullough (1992), 18.

8 "*to observe the interactions of forces*": Knobloch (2007), 6.
"*Natural History must in good time*": L. Agassiz (1859), *An Essay on Classification*, London: Longman, Brown, Green, Longmans, & Roberts.
"*how relatively little had been known*": in McCullough (1992), 17–18.

2. Reverend Darwin's Detour

Thanks to the efforts of many scholars and Darwin's habit of keeping his correspondence, a massive amount of information is available on Darwin's life and work in his diaries, letters, notebooks, and writings. Much of this material can now be found online at http://darwin-online.org.uk/. There are also many fine biographies of Darwin, two of which I drew upon here:

*Browne, Janet (1995). *Charles Darwin: Voyaging*. New York: Alfred A. Knopf.
*Desmond, Adrian, and James Moore (1991). *Darwin: The Life of a Tormented Evolutionist*. London: Michael Joseph.

Various quotes or information have been taken from the following list:

Barlow, Emma Nora (1963). *Darwin's Ornithological Notes*. Bulletin of the British Museum (Natural History) Historical Series, 2: 201–73. (DON)
——— (1967). *Darwin and Henslow: The Growth of an Idea. Letters 1831–1860*. Berkeley: University of California Press. (DH)
The Correspondence of Charles Darwin (1985). F. H. Burkhardt, S. Smith et al., eds. Cambridge, U.K.: Cambridge University Press. (CCD1, CCD2)
Darwin, Charles. Notebooks "B," "D," and "E." Images online at http://darwin-online.org.uk. (NB, ND, NE respectively)
——— (1839). *Journal of Researches into the Geology and Natural History of the Various Countries Visited by H.M.S. Beagle Under the Command of Captain FitzRoy, R. N. from 1832 to 1836*. London: Henry Colbourn. (VB)
Darwin, Erasmus (1803). *Zoonomia: or, The Laws of Organic Life*, Volume 1. Boston: Thomas and Andrews.
Darwin, Francis (ed.) (1887). *The Life and Letters of Charles Darwin, including an autobiographical chapter*, Volumes 1–3. London: John Murray. (LL)
——— (1909). *The Foundations of the Origin of Species. Two Essays Written in 1842 and 1844*. Cambridge, U.K.: Cambridge University Press. (E42 and E44)
Keynes, Richard D. (ed.) (1988). *Charles Darwin's Beagle Diary*. Cambridge, U.K.: Cambridge University Press. (BD)
Lyell, Charles (1832). *The Principles of Geology*, Volume 2. London: John Murray.

SOURCES OF QUOTES

18 *"You care for nothing"*: LL, autobiography, 32.

19 *"He desires me to say that"*: CCD1, 37.

"If this gradual production of species": Zoonomia, 399.

21 *"quite the most perfect man"*: CCD1, 110.

22 *"well-educated and scientific person . . . no opportunity of collecting"*: R. FitzRoy (1839), *Narrative of the Surveying Voyage of the HMS* Adventure *and* Beagle, London: Henry Colbourn, 18.

23 *"I think you are the very man they are in search of"*: DH, 30.

"if you think differently from me": CCD1, 132.

"Disreputable to my character": CCD1, 133.

"all the assistance in my power": CCD1, 135.

25 *"stewed in . . . warm melted butter"*: BD, 35.

"I formerly admired Humboldt": DH, 55.

26 *"We breakfast at eight o'clock"*: CCD1, 248.

"by far the most savage": BD, 99.

27 *"taste & look like a duck"*: BD, 105.

"cargoes of apparent rubbish": see BD, 106 (FitzRoy narrative 2: 106–7).

28 *"I have scarcely for an hour"*: BD, 131.

29 *"May Providence keep"*: BD, 132.

"my messmate, who so willingly": BD, 140 footnote (FitzRoy narrative 2: 216–17).

"I know not, how I shall be able to endure it": DH, 63.

30 *"turned out to be most interesting"*: DH, 77–79.

32 *"I saw the spot where a cluster of fine trees"*: VB, 406–7.

33 *"the stunted trees show little signs of life"*: BD, 352.

34 *"paradise for the whole family of Reptiles"*: BD, 353.

"two very large Tortoises": BD, 354.

"pointing out to me as a youngster": LL1, 225.

36 *"in succession at such times"*: C. Lyell (1832), *Principles of Geology, vol. 2*, London: John Murray, 124.

"I loathe, I abhor the sea": CCD1, 503.

"I have specimens from four": DON, 262.

39 *"[I]t never occurred to me"*: VB, 475.

40 *"one is urged to look to common parent?"*: DON, 277.

all quotes from notebook B from text: at http://darwin-online.org.uk/.

41 *"One may say there is a force"*: ND, 135.

"It is a beautiful part of my theory": NE, 71.

42 *"we can allow satellites, planets, suns"*: P. H. Barrett (1974), "Early writings of Charles Darwin," in *Darwin on Man*, H. E. Gruber, ed. London: Wildwood House, 337.

"the greatest success that my humble work": CCD2, 218–22.

42   *"for the great pleasure"*: CCD2: (letter 545).
44   *"Now the Creationist believes these three Rhinoceroses"*: F. Darwin (1909), *The Foundations of the Origin of Species*, Cambridge: Cambridge University Press, 49.
      "I have just finished my sketch of my species theory": see FOS, xxvi.
45   *"the dear old Philosopher"*: LL1, 221.

3. Drawing a Line between Monkeys and Kangaroos

BOOKS

Beddall, B. G. (1969). *Wallace and Bates in the Tropics: An Introduction to the Theory of Natural Selection*. London: Macmillan Co.
*Quammen, D. (1996). *The Song of the Dodo: Island Biogeography in an Age of Extinction*. New York: Scribner.
van Oosterzee, P. (1997). *Where Worlds Collide: The Wallace Line*. Ithaca, N.Y.: Cornell University Press.
Wallace, A. R. (1890). *The Malay Archipelago*. London: Macmillan and Co.
———— (1905). *My Life*. New York: Dodd, Mead, and Co.

ARTICLES

Forbes, H. O. (1914). Obituary: Alfred Russel Wallace, O.M. *Geographical Journal* 43: 88–92.
McKinney, H. Lewis (1969). Wallace's earliest observations on evolution. *Isis* 60: 370–73.
Wallace, A. R. (1855). On the law which has regulated the introduction of new species. *Annals and Magazine of Natural History* 16: 184–96.
———— (1857). On the natural history of the Aru Islands. *Annals and Magazine of Natural History* 20: 473–85.
———— (1858). On the tendency of varieties to depart indefinitely from the original type. *Proceedings of the Linnean Society of London* 3: 53–62.

SOURCES OF QUOTES

48   *"I'm afraid the ship's on fire"*: My Life, 303–12.
      "It was now, when the danger": ibid.
50   *"Here and there, too, were tiger pits"*: Malay Archipelago, 18.
51   *"It was rather nervous work hunting for insects"*: ibid., 18–19.
      "as there were many bad people about": in A. R. Wallace (1856), Letter from Macassar, Celebes, *Zoologist* 15: 5559–60.
52   *"Nature seems to have taken every precaution"*: My Life, 394.
      "I had seen sitting on a leaf": Malay Archipelago, 257–58.

53 *"Every species has come into existence"*: Wallace (1855).

54 *"that the present geographical distribution of life"*: ibid.
"which contain little groups of plants and animals": ibid.
"They must have been first peopled": ibid.
"They could not be as they are": ibid.
"like truth itself, so simple and obvious": J. Marchant (1916), *Alfred Russel Wallace: Letters and Reminiscences, vol. 1.* London: Cassell and Company, Ltd.

55 *"crossing over to Lombok"*: *Malay Archipelago*, 155.

56 *"Let us now examine"*: Wallace (1857).
"We can scarcely find a stronger contrast": ibid.
"some other law has regulated the distribution of existing species": ibid.

57 *"I agree to the truth"*: Darwin Correspondence Project, http://www.darwin project.ac.uk/ Letter 2086, C. R. Darwin to A. R. Wallace, May 1857.
"to think over subjects then particularly interesting to me": *My Life*, 361.

58 *"The life of wild animals is a struggle for existence"*: Wallace (1858).

59 *"I know not how or to whom"*: Wallace letter to H. W. Bates, Dec. 1860, Wallace Collection, Natural History Museum, Online Transcription, accessed 3/12/08. http://www.nhm.ac.uk/nature-online/collections -at-the-museum/wallace-collection/item.jsp?itemID=70&theme=/.
"I am a little proud": *My Life*, 366.

4. *Life Imitates Life*

BOOKS

Bates, H. W. (1892). *The Naturalist on the River Amazons,* with a memoir of the author by Edward Clodd. London: John Murray. (NORA)

The Correspondence of Charles Darwin, Volume 9, 1861. Cambridge, U.K.: Cambridge University Press.

The Correspondence of Charles Darwin, Volume 10, 1862. Cambridge, U.K.: Cambridge University Press.

ARTICLES

Allen, G. (1862). Bates of the Amazons. *Fortnightly Review* 58: 798–809.

Bates, H. W. (1862). Contributions to an insect fauna of the Amazon Valley. Lepidoptera: *Helaconidae. Transactions of the Linnean Society* 23: 495–566. (TLS)

Brower, J.V.Z. (1958). Experimental studies of mimicry in some North American butterflies. Part I. The Monarch, *Danaus flexippus,* and Viceroy, *Limenitis arcippus archippus. Evolution* 12: 32–47.

Brower, L. P., J.V.Z. Brawer, and C. T. Collins (1963). Experimental studies of mimicry. Relative palatability and Müllerian mimicry among neotropical butterflies of the subfamily *Heliconiinae. Zoologica* 48: 65–83.

Darwin, C. D. (1863). A review of Mr. Bates' paper on "Mimetic Butterflies." *Natural History Review,* 219–24.

O'Hara, J. E. (1995). Henry Walter Bates—his life and contributions to biology. *Archives of Natural History* 22: 195–219.

Pfennig, D., W. R. Harcombe, and K. S. Pfenning (2001). Frequency-dependent Batesian mimicry. *Nature* 410: 323.

Wallace, A. R. (1866). Natural Selection. *Athenaeum,* December 1, no. 2040: 716–17.

SOURCES OF QUOTES

61   "*I rose generally with the sun*": NORA, 269.
     "*Twelve months elapsed*": ibid., 271.

63   "*On the evening of the third*": ibid., 388–89.
     "*I think I have got a glimpse into the laboratory*": *Correspondence,* Vol. 9, 74.

65   *accidental resemblances:* TLS, 510.

67   "*The explanation of this*": ibid., 511.

68   "*one of the most remarkable*": *Correspondence,* Vol. 10 (November 20, 1862).
     "*turned out all ready made . . . on this earth*": Darwin (1863), 219–24.

70   "*My criticisms may be condensed*": *Correspondence,* Vol. 11 (April 18, 1863).

71   "*It may be said, therefore*": NORA, 353.

5. Java Man

BOOKS

Darwin, Charles (1871). *The Descent of Man and Selection in Relation to Sex.* London: John Murray.

Haeckel, Ernst (1887). *The History of Creation: Or the Development of the Earth and Its Inhabitants by the Action of Natural Causes.* Translation revised by E. Ray Lankester. New York: Appleton Company.

Huxley, Thomas H. (1959). *Man's Place in Nature.* Ann Arbor: University of Michigan Press (reprint of 1863 edition).

*Shipman, Pat (2001). *The Man Who Found the Missing Link: Eugène Dubois and His Lifelong Quest to Prove Darwin Right.* New York: Simon and Schuster.

Theunissen, Bert (1989). *Eugène Dubois and the Ape-Man from Java: The History of the First "Missing Link" and Its Discoverer.* Dardrecht, The Netherlands: Kluwer Academic Publishers.

ARTICLES

de Vos, John (2004). "The Dubois collection: a new look at an old collection." In *VII International Symposium Cultural Heritage in Geosciences, Mining, and Metallurgy: Libraries—Archives—Museums: Museums and their Collections. Scipla Geologic,* Special Issue 4: 267–85.

SOURCES OF QUOTES

82 *"The question of questions for mankind":* Huxley, 71.
    "Let us . . . disconnect our thinking selves": ibid., 85.
83 *"is man so different":* ibid., 85–86.
    "Being happily free from all real": ibid., 86.
84 "the two pairs of limbs and": Haeckel, 299.
    "Speechless Man (Alalus)": ibid., 300.
    "Where, then, must we look for primaevel Man?": Huxley, 184.
89 *"Everything here has gone against me":* Theunissen, 40.
92 *"I discover that there is no more unsuitable place":* Shipman, 159.
94 "Pithecanthropus erectus *is the transitional form":* Eugène Dubois (1894), *Pithecanthropus erectus:* Eine menschen-aehnliche Vebergangstorm aus Java, *Batavia* (cited in Shipman, 209).
97 *"able discoverer of* Pithecanthropus*":* Ernst Haeckel (1898), "On our present knowledge of the Origin of Man," *Annual Report of the Board of Regents of the Smithsonian Institution for the Year Ending June 30, 1898,* translation of a discourse given at the Fourth International Congress of Zoology at Cambridge, England (cited in Shipman, 306–7).

6. To the Big Bang, on Horseback

BOOKS

Darwin, C. R. (1869). *On the Origin of Species by Means of Natural Selection, or The Preservation of Favoured Races in the Struggle for Life.* London: John Murray, 5th edition.

Gould, S. (1989). *Wonderful Life: The Burgess Shale and the Nature of History.* New York: W. W. Norton.

Hou, X., et al. (2004). *The Cambrian Fossils of Chengjiang, China: The Flowering of Early Animal Life.* Malden, Mass.: Blackwell.

Knoll, A. (2005). *Life on a Young Planet: The First Three Billion Years of Evolution on Earth.* Princeton, N.J.: Princeton University Press.

Morris, S. C. (2000). *The Crucible of Creation: The Burgess Shale and the Rise of Animals.* London: Oxford University Press.

Schopf, J. (1999). *Cradle of Life: The Discovery of Earth's Earliest Fossils.* Princeton, N.J.: Princeton University Press.

Yochelson, E. (1998). *Charles Doolittle Walcott, Paleontologist.* Kent, Ohio: Kent State University Press.

———. (2001). *Smithsonian Institution Secretary, Charles Doolittle Walcott.* Kent, Ohio: Kent State University Press.

ARTICLES

Morris, S. C. (2006). Darwin's dilemma: The realities of the Cambrian 'explosion.' *Philosophical Transactions of the Royal Society of London B* 361: 1069–83.

Schopf, J. W. (2000). Solution to Darwin's dilemma: Discovery of the Missing Precambrian record of life. *Proceedings of the National Academy of Sciences* 97: 6947–53.

Walcott, C. D. (1883). Pre-Carboniferous Strata in the Grand Cañon of the Colorado, Arizona. *American Journal of Science* 26: 437–42.

——— (1884). Report of Mr. Charles D. Walcott. *Fourth annual report of the United States Geological Survey to the Secretary of the Interior 1882–83*: 44–48.

——— (1891). "The North American Continent During Cambrian time." *Twelfth Annual Report of the United States Geological Society to the Secretary of the Interior 1890–1891*, 12: 526–68.

——— (1899). Pre-Cambrian fossiliferous formations. *Bulletin of the Geological Society of America* 10: 199–214.

——— (1911). Cambrian Geology and Paleontology: Abrupt appearance of the Cambrian fauna on the North American continent. *Smithsonian Miscellaneous Collections* 57: 1–16.

——— (1911). Cambrian Geology and Paleontology: Middle Cambrian Merostomata. *Smithsonian Miscellaneous Collections* 57: 17–40.

——— (1911). A Geologist's Paradise. *National Geographic Magazine* 22: 509–36.

Yochelson, E. L. (1967). Charles Doolittle Walcott: March 31, 1850–February 9, 1927. *National Academy of Sciences Biographical Memoirs* 39: 470–540.

——— (2006). Charles D. Walcott: A few comments on Stratigraphy and Sedimentation. *Sedimentary Record* 4: 4–8.

WEBSITES

"Beauty in Service to Science: Treasures of the Burgess Shale." 3 November 2004. Smithsonian Institution Archives. 21 November 2006. http://siarchives.si.edu/techsvcs/walcott/treasures.htm

Lee, Ronald F. "The Story of the Antiquities Act: Chapter 6, The Antiquities Act, 1900–1906." 2001. *NPS Archaeology Program: For the Public.* National Park Service U.S. Department of the Interior. http://www .nps.gov/history/archeology/pubs/Lee/Lee_CH6.htm

"A Voice from the Cambrian." 1 August 2007. Smithsonian. 24 November 2007. http://www.150.si.edu/chap7/seven.htm

SOURCES OF QUOTES

103   *"the finest known from the Trenton group":* in Yochelson (1998), 31.

104   *"I looked to Professor Agassiz as a guide":* ibid., 33.

105   *"the most tedious disagreeable ride":* ibid., 90.

106   *"The view from the summit of the White Cliffs":* ibid.
         "Thus far I have enjoyed my trip very much": ibid., 91.

108   *"Last summer a horse trail was built":* ibid., 151.
         "There is another . . . difficulty, which is much more serious": Darwin (1869), 378.

110   *"may be truly urged as a valid argument":* ibid., 381.
         "ripple marks and mud cracks abound": Walcott (1883), 441.
         "But for the discovery of a small Discinoid": ibid.
         "could scarcely have been the only representations": ibid.

111   *"otherwise known as fieldwork":* Yochelson (1998), 341.

112   *"Walcott, you may have a building for the survey . . .":* ibid., 213.

113   *"tested and tried and we know how well":* ibid., 399.

114   *"a man who was not only a recognized scientist":* ibid., 463.
         "Now it seems to me that this opens the best chance": A Voice from the Cambrian.

116   *"The collection is great":* in Yochelson (2005), 69.

119   *"Those 50 million years reshaped more than":* Knoll (2005), 193.
         "steady, systematic work is one's salvation": in Yochelson (2001), 185–86.

7. *Where the Dragon Laid Her Eggs*

BOOKS

*Andrews, Roy Chapman (1926). *On the Trail of Ancient Man: A Narrative of the Field Work of the Central Asiatic Expeditions.* New York: Putnam.

——— (1932). *The New Conquest of Central Asia; a Narrative of the Explorations of the Central Asiatic Expeditions in Mongolia and China, 1921–1930.* New York: American Museum of Natural History.

——— (1943) *Under a Lucky Star: A Lifetime of Adventure.* New York: Viking Press.

Bausum, Ann. (2000) *Dragon Bones and Dinosaur Eggs: A Photobiography of Explorer Roy Chapman Andrews.* Washington, D.C.: National Geographic Society.

Gallenkamp, Charles (2001). *Dragon Hunter: Roy Chapman Andrews and the Central Asiatic Expeditions.* New York: Viking.

Rexer, Lyle (1995). *American Museum of Natural History: 125 Years of Expedition and Discovery.* New York: H. N. Abrams in association with the American Museum of Natural History.

SOURCES OF QUOTES

124   *"so you can get 'em all next time"*: Andrews (1943), *Under a Lucky Star*, 12.

125   *"I'm not asking for a position"*: ibid., 22.

126   *"At last, there it was spreading its length like a slumbering gray serpent"*: ibid., 116.

127   *"Never again will I have such a feeling as Mongolia gave me"*: in C. Gallenkamp (2001), *Dragon Hunter*, 73.

128   *"should try to reconstruct the whole past history of the Central Asian plateau"*: Andrews (1943), *Under a Lucky Star*, 163.
       "Roy, we've got to do it. This plan": ibid., 164.
       "It's a great plan; a great plan": ibid., 167.

130   *"geology was all obscured by sand"*: Andrews (1932), *New Conquest of Central Asia*, 7.

131   *"Well, Roy, we've done it. The stuff is here"*: Andrews (1926), *On the Trail of Ancient Man*, 78.

132   *"Slowly I became conscious that the air was vibrating"*: Andrews (1943), *Under a Lucky Star*, 191.

134   *"were paved with white fossil bones and all represented animals unknown"*: Andrews (1926), *On the Trail of Ancient Man*, 216.

135   *"You have written a new chapter"*: Andrews (1943), *Under a Lucky Star*, 200.
       "great fun": Andrews (1926), *On the Trail of Ancient Man*, 216.

135 *"Then our indifference suddenly evaporated"*: Andrews (1943), *Under a Lucky Star*, 213, and Andrews (1926), *On the Trail of Ancient Man*, 228–29.

136 *"Nothing in the world was further from our minds"*: Andrews (1943), *Under a Lucky Star*, 215.

"While the rest of us were on our hands and knees": ibid., 213.

137 *"Let's keep the flour for work"*: Andrews (1926), *On the Trail of Ancient Man*, 232.

"an unidentifiable reptile": ibid., 327.

"Do your utmost to get some other skulls": ibid., 328.

"Well, I guess that's an order": ibid.

139 *"possibly the most valuable seven days of work"*: ibid., 329.

140 *"Dear God, my tent is full of snakes"*: Andrews (1935), *This Business of Exploring*, New York: G. P. Putnam's Sons, 41.

"fight its way across the long miles of desert": Andrews (1932), *New Conquest of Central Asia*, 310.

141 *"always intended to be an explorer"*: Andrews (1943), *Under a Lucky Star*, 13.

"always there has been an adventure": ibid., 300.

8. The Day the Mesozoic Died

BOOKS

Alvarez, Luis W. (1987). *Alvarez: Adventures of a Physicist.* New York: Basic Books.

*Alvarez, Walter (1997). *T. rex and the Crater of Doom.* Princeton, N.J.: Princeton University Press.

Powell, James L. (1998). *Night Comes to the Cretaceous: Dinosaur Extinction and the Transformation of Modern Geology.* New York: Freeman.

ARTICLES

Alvarez, L. (1983). Experimental evidence that an asteroid impact led to the extinction of many species 65 million years ago. *Proceedings of the National Academy of Sciences* 80: 627–42.

Alvarez, L., et al. (1980). Extraterrestrial Cause for the Cretaceous-Tertiary Extinction: Experimental results and theoretical interpretation. *Science* 208: 1095–1108.

Alvarez, W., et al. (1990). Iridium Profile for 10 Million Years Across the Cretaceous-Tertiary Boundary at Gubbio (Italy). *Science* 250: 1700–1702.

Claeys, P., et al. (2002). Distribution of Chicxulub ejecta at the Cretaceous-Tertiary boundary. *Geological Society of America*, Special Paper 356: 55–68.

Hildebrand, A. R. (1991). Chicxulub Crater: A possible Cretaceous/Tertiary boundary impact crater on the Yucatán Peninsula, Mexico. *Geology* 19: 867–71.

Kring, D., and D. Durda (2003). The Day the World Burned. *Scientific American* 289: 98–105.

Mukhopadhyay, S., et al. (2001). A Short Duration of the Cretaceous-Tertiary Boundary Event: Evidence from Extraterrestrial Helium-3. *Science* 291: 1952–55.

Orth, C. J., et al. (1981). An Iridium Abundance Anomaly at the Palynological Cretaceous-Tertiary Boundary in Northern New Mexico. *Science* 214: 1341–43.

Pope, K. O. (1991). Mexican site for K/T impact crater? *Nature*, Scientific Correspondence 351: 105.

Pope, K. O., et al. (1998). Meteorite impact and the mass extinction of species at the Cretaceous/Tertiary Boundary." *Proceedings of the National Academy of Sciences* 95: 11028–29.

Schuraytz, B. C., et al. (1996). Iridium Metal in Chicxulub Impact Melt: Forensic Chemistry on the K-T Smoking Gun. *Science* 271: 1573–76.

Simonson, B. M., and B. P. Glass (2004). Spherule Layers—Records of Ancient Impacts. *Annual Review of Earth Planet Sciences* 32: 329–31.

Smit, J. (1999). The Global Stratigraphy of the Cretaceous-Tertiary Boundary Impact Ejecta. *Annual Review of Earth Planet Sciences* 27: 75–113.

WEBSITES

"Luis Alvarez: The Nobel Prize in Physics 1968, Biography." From Nobel Lectures, Physics 1963–1970, Elsevier Publishing Company, Amsterdam, 1972. http://nobelprize.org/nobel_prizes/physics/laureates/1968/alvarez-bio.html

"Tiny Creatures Tell a Big Story: Blast from the Past!—Part 2." Department of Paleobiology, Smithsonian National Museum of Natural History. http://paleobiology.si.edu/blastPast/paleoBlast2.html

SOURCE OF QUOTE

150 *"something unpleasant had happened to the Danish sea bottom"*: Alvarez (1997), *T. Rex and the Crater of Doom*, 70.

9. Dinosaurs of a Feather

BOOKS

*Chiappe, Luis M. (2007). *Glorified Dinosaurs: The Origin and Early Evolution of Birds.* Hoboken, N.J.: John Wiley.

Dingus, Lowell, and Timothy Rowe (1998). *The Mistaken Extinction: Dinosaur Evolution and the Origin of Birds.* New York: W. H. Freeman.

*Shipman, Pat (1998). *Taking Wing: Archaeopteryx and the Evolution of Bird Flight.* New York: Simon & Schuster.

Wilford, John Noble (1985). *The Riddle of the Dinosaur.* New York: Alfred A. Knopf.

ARTICLES

Browne, Malcolm W. (October 19, 1996). "Feathery Fossil Hints Dinosaur-Bird Link." *New York Times,* Section 1; Page 1; Column 3.

———. (April 25, 1997). "In China, a Spectacular Trove of Dinosaur Fossils Is Found." *New York Times,* Section 1; Page 1; Column 3.

Cracraft, Joel (1977). Special Review: John Ostrom's Studies on Archaeopteryx, the origin of birds, and the evolution of avian flight. *Wilson Bulletin* 89.3: 488–92.

Darwin Correspondence Project (January 3, 1863). "Letter 3899—Falconer, Hugh to Darwin, C. R., 3 Jan [1863]." http://www.darwinproject.ac.uk/darwinletters/calendar/entry-3899.html

——— (January 5, 1863). "Letter 3901—Darwin, C. R. to Falconer, Hugh, 5 [and 6] Jan [1863]." http://www.darwinproject.ac.uk/darwinletters/calendar/entry-3901.html

——— (January 7, 1863). "Letter 3905—Darwin, C. R. to Dana, J. D., 7 Jan [1863]." http://www.darwinproject.ac.uk/darwinletters/calendar/entry-3905.html

——— (January 8, 1863). "Letter 3908—Falconer, Hugh to Darwin, C. R., 8 Jan [1863]." http://www.darwinproject.ac.uk/darwinletters/calendar/entry-3908.html

Hecht, Jeff (July 21, 2005). "Obituary: Professor John Ostrom: Palaeontologist who showed that birds were descended from dinosaurs." *Independent (London),* Obituaries, 52.

Huxley, Thomas H. (1868). On the animals which are most nearly intermediate between birds and reptiles. *Geological Magazine* 5: 357–65.

Monastersky, R. (October 26, 1996). "Hints of a Downy Dinosaur in China." *Science News Online.* http://www.sciencenews.org/pages/sn_arch/10_26_96/fob1.htm

Musante, Fred (June 29, 1997). "Connecticut Q&A: Dr. John H. Ostrom; Lessons for the Future in Ancient Bones." *New York Times*, Section 13CN; Page 3; Column 1.

NPR (National Public Radio) (July 21, 2005). "Influential Paleontologist John Ostrom, 77, Dies." *All Things Considered*.

Ostrom, John H. (1969). "A New Theropod Dinosaur from the lower Cretaceous of Montana." *Postilla*, New Haven, Conn., Peabody Museum of Natural History, 128: 1–17.

———— (1969). "Osteology of Deinonychus antirrhopus, An Unusual Theropod from the Lower Cretaceous of Montana." *Peabody Museum of Natural History* Bulletin 30, 1–164.

———— (1969). The Supporting Chain. *Discovery* 5: 10–16.

———— (1969). Terrible Claw. *Discovery* 5: 1–9.

———— (1970). Archaeopteryx: Notice of a 'New' Specimen. *Science* 170: 537.

———— (1973). The Ancestry of Birds. *Nature* 242: 136.

———— (1975). Archaeopteryx. *Discovery* 11.1: 15–23.

———— (1975). The Origin of Birds. *Annual Review of Earth Planet Sciences* 1975: 55–77.

———— (1978). A new look at dinosaurs. *National Geographic* 154: 152–85.

Owen, R. (1842). "Report on British Fossil Reptiles." Report of the Eleventh Meeting of the British Association for the Advancement of Science, held at Plymouth, England (1841), 60–204.

———— (1863). "On the Archaeopteryx of Von Meyer, with a description of the fossil remains of a long-tailed species from the lithographic stone of Solnhofen." *Philosophical Transactions of the Royal Society of London* 153: 33–47.

Special to the *New York Times* (December 4, 1964). "A New Species of Small Dinosaur Reported Found by Yale Curator." *New York Times*.

Spielberg, S. (director) (1994). *Jurassic Park*. Universal City, Calif.: Universal Pictures, Amblin Entertainment.

Torrens, Hugh (April 4, 1992). When did the dinosaur get its name?: The name dinosaur was coined 150 years ago last year—or was it this year? Clever detective work has solved the puzzle. *New Scientist* 1815: 40.

Wilford, John Noble (July 21, 2005). "John H. Ostrom, Influential Paleontologist, Is Dead at 77." *New York Times*, Section A; Column 1; 27.

Zimmer, Carl (1992). Ruffled feathers: A paleontologist going after the earliest bird may have ended up with a mouthful of worms. *Discover* 13.5: 4–54.

SOURCES OF QUOTES

162 *"I was almost certain, although still wary"*: Ostrom (1969), *Discovery* 5 (1): 11.

164 *"As by this theory innumerable"*: Darwin (1859), *On the Origin of Species*, London: Murray, 172.

"geology assuredly does not reveal": ibid., 280.

"the abrupt manner in which whole groups": ibid., 302.

167 *"unequivocally to be a Bird"*: Richard Owen (1863), 46.

"You were never more missed": Falconer letter to Darwin, January 3, 1863.

"You are not to put your faith": ibid.

"sort of misbegotten-bird-creature": ibid.

"a grand case for me": Darwin letter to J. D. Dana, January 7, 1863.

168 *"The period when the class of Reptiles flourished"*: Owen (1842), 201.

"the different species of Reptiles were suddenly introduced upon the earth's surface": ibid., 202.

"Dirty Dick": Falconer letter to Darwin, January 8, 1863.

"It is admitted on all sides": Huxley (1868), 304.

169 *"Thus it is a matter of fact"*: ibid., 312.

"Surely, there is nothing very wild or illegitimate": ibid.

172 *"Reptiles are just not capable of such maneuvers"*: Ostrom (1969), *Discovery* 5 (1): 4.

173 *"You blew it, John. You blew it!"*. . . *"Here, here, Professor Ostrom"*: Shipman (1998), 42.

174 *"Whoa, wait a minute"*: Zimmer (1992), 46.

"Indeed, if feather impressions": Ostrom (1973), *Nature* 242: 136.

175 *"full of crap"*: Zimmer (1992), 48.

176 *"in a state of shock"*: Browne (1997), 1.

"weak in the knees": Wilford (2005), 27.

"This was one of the most exciting": Browne (1997), 1.

"As we discover more and more theropod": Chiappe (2007), 17.

178 *"Professor Ostrom, this is Michael Crichton"*: Musante (1997), 3.

179 *"No wonder these guys learned to fly"*: *Jurassic Park*.

"You really think dinosaurs turned into birds?": ibid.

10. It's a Fishapod!

BOOKS

*Shubin, N. (2008). *Your Inner Fish: A Journey into the 3.5-Billion-Year History of the Human Body.* New York: Pantheon Books.

ARTICLES

Ahlberg, P. E., and J. A. Clack (2006). A firm step from water to land. *Nature* 440: 747–49.

Daeschler, E. B., N. H. Shubin, et al. (2006). A Devonian tetrapod-like fish and the evolution of the tetrapod body plan. *Nature* 440: 757–63.

Embry, A., and J. E. Klovan (1976). The Middle-Upper Devonian Clastic Wedge of the Franklinian Geosyncline. *Bulletin of Canadian Petroleum Geology* 24: 485–639.

Shubin, N. H., E. B. Daeschler, et al. (2006). The pectoral fin of Tiktaalik roseae and the origin of the tetrapod limb. *Nature* 440: 764–71.

WEBSITES

Murphy, D. C. "Devonian Times." 2005–2006. October 24, 2007. http://www.devoniantimes.org/

"Tiktaalik roseae." University of Chicago. 2006–2007. October 24, 2007. http://tiktaalik.uchicago.edu/

SOURCES OF QUOTES

185 *"You will soon be at the top of the world":* Interview with Neil Shubin, Chicago, October 9, 2007.

186 *"the . . . fossil content of the Fram Fn . . .* similar to the Catskill Fn of Pennsylvania": Embry and Klovan, 548.

187 *"Great idea. You are going to find":* Interview with Neil Shubin, Chicago, ibid., and phone interview, October 30, 2007.

188 *"Oh, you are not really going* there *are you? . . . It was like saying we were going to Castle Dracula":* Interview with Neil Shubin, Chicago.

197 *"When people call* Tiktaalik *'the missing link'":* L. Helmuth, *Smithsonian,* June 2006, interview with Neil Shubin, at http://www.smithsonianmag.com/science-nature/interview-shubin.html.

11. Journey to the Stone Age

BOOKS

Cole, Sonia (1973). *Leakey's Luck: The Life of Louis Seymour Bazett Leakey, 1903–1972*. New York: Harcourt Brace Jovanovich.

Leakey, L.S.B. (1966). *White African: An Early Autobiography*. Cambridge, Mass.: Schenkman.

————. (1974). *By the Evidence: Memoirs, 1932–1951*. New York: Harcourt Brace Jovanovich.

Leakey, Mary (1984). *Disclosing the Past*. Garden City, N.Y.: Doubleday.

Leakey, Richard, and Roger Lewin (1992). *Origins Reconsidered: In Search of What Makes Us Human*. New York: Doubleday.

*Morrell, Virginia (1995). *Ancestral Passions: The Leakey Family and the Quest for Humankind's Beginnings*. New York: Simon and Schuster.

To get a better appreciation of the process and the effort involved in the Olduvai excavations, and the massive quantities of tools and bones that were recovered, it is well worth perusing the series of volumes produced by the Leakeys:

Leakey, L.S.B. (1951). *Olduvai Gorge 1951–1961, Volume 1*, Cambridge: Cambridge University Press.

———— (1965). *Olduvai Gorge: A Report on the Evolution of the Hand-axe Culture in Beds I–IV*. Cambridge, U.K.: Cambridge University Press.

*Leakey, Mary D. (1971). *Olduvai Gorge: Volume 3, Excavations in Beds I and II, 1960–1963*. Cambridge, U.K.: Cambridge University Press.

Leakey, Mary D., and Derik R. Roe (1994). *Olduvai Gorge, Volume 5, Excavations in Beds III, IV, and the Masek Beds, 1968–1971*. Cambridge: Cambridge University Press.

Essential scientific articles documenting some of the findings described include:

Leakey, L.S.B. (1959). *Nature* 184: 491–93. (The discovery of Dear Boy)

Leakey, L.S.B., J. F. Evernden, and G. H. Curtis (1961). *Nature* 191: 478–79. (Age of Olduvai Bed I)

Leakey, L.S.B, P. V. Tobias, and J. R. Napier (1964). *Nature* 202: 7–9. (Discovery of *Homo habilis*)

Leakey, M. D. (1966). *Nature* 210: 462–66. (Olduwan tools)

Leakey, M. D., and R. L. Hay (1979). *Nature* 278: 317–23. (Laetoli footprints)

Bye, B. A., F. H. Brown, T. E. Cerling, and I. McDougal (1987). *Nature* 329: 237–39. (Age of Olorgesailie)

OTHER ARTICLES

Berkey, C. P. (1929). *Scientific Monthly* 28: 193–216 (for a geologist's conception of time in the late 1920s).

Leakey, M. D. (1981). *Philosophical Transactions of the Royal Society of London B.* 292: 95–102 (for more on tracks and tools).

Susman, R. L. (1991). *Journal of Anthropological Research* 47: 129–51 (on who made the Olduvai tools).

Wood, B. (1989). *Current Anthropology* 30: 215–24 (for insights into working with Louis Leakey).

SOURCES OF QUOTES

207 *"I did everything except feed"*: *White African*, 33.

210 *"a great adventure"*: ibid., 107.

211 *"a strange mixed feeling"*: ibid., 114.
    "I little thought when I was kicked in the head": ibid., 97.

212 *"It is somewhat more probable"*: Charles Darwin (1871), *The Descent of Man and Selection in Relation to Sex*, Penguin Classics edition (2004), 182.
    "everyone knew he had started in Asia": *White African*, preface to 1966 edition.

215 *"contained a liquid consisting"*: ibid., 287.

216 *"a scientist's paradise"*: ibid., 296.

217 *"a very remote period indeed"*: J. Frere (1800), *Achaeologia* 13: 204–5.
    "At least I ended my school career": *Disclosing the Past*, 33.
    "When we finally reached the top": ibid., 54–55.

218 *"I still am convinced"*: Cited in *Ancestral Passions*, 98.

219 *"She and I were mutually horrified"*: *Disclosing the Past*, 59.
    "cast a spell on her": ibid., 63.
    After an exhilarating: Richard Leakey (1983), *One Life: An Autobiography*, London: Michael Joseph, 23.
    "the most complete record of a tribe": Cited in *Ancestral Passions*, 113.

220 *"When I saw her site"*: *By the Evidence*, 159–60.

222 *"never cared in the least for crocodiles"*: *Disclosing the Past*, 98.
    "that night we cast aside care": ibid., 99.

225 *"Him, the man! Our man"*: L.S.B. Leakey, *National Geographic*, September 1960, 431.
    "quite lovely": *Ancestral Passions*, 183.

226 *"It absolutely sent shivers down my spine"*: ibid., 187.
    "I am so glad that this has happened": Wood (1989), 217.

229 *"Oh, then I think it must be a hominid"*: *Disclosing the Past*, 126.
    "Come quickly, top secret, we've got the Man": Wood (1989), 219.
    "hair stood on end": *Leakey's Luck*, 253.

233 *"Now this is really something"*: John Reader (1981), *Missing Links: The Hunt for Earliest Man*, Boston: Little, Brown and Company, 15.

12. Clocks, Trees, and H-Bombs

BOOKS

Hager, Thomas (1995). *Force of Nature: The Life of Linus Pauling.* New York: Simon and Schuster.

Leakey, Richard, and Roger Lewin (1992). *Origins Reconsidered: In Search of What Makes Us Human.* New York: Doubleday.

Lewin, Roger (1997). *Bones of Contention,* 2nd ed. Chicago: University of Chicago Press.

ARTICLES

The Clock

Morgan, G. J. (1998). *Journal of the History of Biology* 31: 155–78.

Simpson, G. G. (1964). *Science* 146: 1535–38.

Zuckerkandl, E., R. T. Jones, and L. Pauling (1960). *Proceedings of the National Academy of Sciences USA* 46: 1349–60.

Zuckerkandl, E., and W. A. Schroeder (1961). *Nature* 192: 984–85.

Zuckerkandl, E., and L. Pauling (1962). "Molecular Disease, Evolution, and Genetic Heterogeneity," in *Horizons in Biochemistry: Albert Szent-Györgi Dedicatory Volume,* M. Kasha and B. Pullman, eds. New York: Academic Press, 189–225.

————— (1965). "Evolutionary Divergence and Convergence in Proteins," in *Evolving Genes and Proteins,* V. Bryson and H. Vogel, eds. New York: Academic Press, 97–166.

Dating the Chimpanzee-Human Split

Sarich, V. M. (1971). "A Molecular Approach to the Question of Human Origins," in *Background for Man,* P. J. Dolhinow and V. M. Sarich, eds. Boston: Little, Brown and Company, 60–81.

Sarich, V. M., and A. C. Wilson (1966). *Science* 154: 1563–66.

————— (1967). *Science* 158: 1200–1203.

————— (1967). *Proceedings of the National Academy of Sciences USA* 58: 142–48.

Wilson, A. C., and V. M. Sarich (1969). *Proceedings of the National Academy of Sciences USA* 63: 1088–93.

Criticism

Buettner-Janusch, J. (1968). *Transactions of the New York Academy of Sciences* 31: 128–38.

Leakey, L.S.B. (1970). *Proceedings of the National Academy of Sciences USA* 67: 746–48.

Simons, E. (1968). *Annals of the New York Academy of Sciences* 167: 319–31.

Resolution

Andrews, P., and J. E. Cronin (1982). *Nature* 297: 541–46.

Atomic Weapons, Test Ban Treaty, and Peace Prize

Nobel Lecture: http:/-nobelprize.org/nobel_prizes/peace/laureates/1962
/pauling-lecture.html.

The Nobel Peace Prize 1962 Presentation Speech: http:/-nobelprize.org
/nobel_prizes/peace/laureates/1962/press.html

SOURCES OF QUOTES

238   *Medal of Merit citation: Ava Helen and Linus Pauling Papers*, Oregon
State University Special Collections.

240   *"What is your reaction":* Hager, *Force of Nature*, 450.
"I must emphasize": ibid., 451.

241   *The petition: Bulletin of the Atomic Scientists* (1957) 13: 264–66.

243   *"something outrageous":* Morgan (1998), 164.
"the number of differences": Zuckerkandl and Pauling (1962), 201.
"falls on the lower limit": ibid., 202.

244   *"bipedal, tool-making":* E. Mayr (1963), in *Classification and Evolution*,
Sherwood Washburn, ed. (Aldine, Chicago), 344.
"it will be possible": Morgan (1998), 168.

245   *Telegram to Kennedy: Ava Helen and Linus Pauling Papers*, Oregon State
University Special Collections.

246   *"Daddy, have you heard":* Hager, *Force of Nature*, 546.

248   *"There is a strong consensus":* Simpson (1964).
"There may thus exist": Zuckerkandl and Pauling (1965), 148.

250   *"entirely contrary":* Leakey (1970).
"Students of human origins": Simons (1968), 328.

253   *"If Sarich and Wilson":* Buettner-Janusch (1968), 133.
"I am not a biochemist": Simons (1968), 326.
"one no longer": Sarich (1971), 76.

255   *"I am staggered":* Leakey and Lewin (1992), 78.

13. CSI: Neander Valley

Mitochondrial DNA, Eve, and Allan Wilson

Cann, R. L. (1993). Obituary: Allan C. Wilson, 1935–1991. *Human Biology*,
343.

Cann, R. L., M. Stoneking, et al. (1987). Mitochondrial DNA and human
evolution. *Nature* 325: 31–36.

Higuchi, R., B. Bowman, et al. (1984). DNA sequences from the quagga, an extinct member of the horse family. *Nature* 312: 282–84.

Prugnolle, F., A. Manica, et al. (2005). Geography predicts neutral genetic diversity of human populations. *Current Biology* 15: R159-60.

Vigilant, L., M. Stoneking, et al. (1991). African Populations and the Evolution of Human Mitochondrial DNA. *Science* 253: 1503–7.

Ancient DNA and Neanderthals

Caramelli, D., C. Lalueza-Fox, et al. (2003). Evidence for a genetic discontinuity between Neandertals and 24,000-year-old anatomically modern Europeans. *Proceedings of the National Academy of Sciences USA* 100: 6593–97.

Green, R. E., J. Krause, et al. (2006). Analysis of one million base pairs of Neanderthal DNA. *Nature* 444: 330–36.

Krause, J., L. Orlando, et al. (2007). Neanderthals in central Asia and Siberia. *Nature* 449: 902–4.

Krings, M., H. Geisert, R. Schmitz, H. Krainitzki, and S. Pääbo (1999). DNA sequence of the mitochondrial hypervariable region II from the Neandertal type specimen. *Proceedings of the National Academy of Sciences USA* 96: 5581–85.

Krings, M., A. Stone, et al. (1997). Neandertal DNA Sequences and the Origin of Modern Humans. *Cell* 90: 19–30.

Lindahl, T. (1997). Facts and Artifacts of Ancient DNA. *Cell* 90: 1–3.

Noonan, J. P., G. Coop, et al. (2006). Sequencing and Analysis of Neanderthal Genomic DNA. *Science* 314: 1113–18.

Pääbo, S. (1985). Molecular cloning of Ancient Egyptian mummy DNA. *Nature* 314: 644–45.

Serre, D., A. Langaney, et al. (2004). No Evidence of Neandertal mtDNA Contribution to Early Modern Humans. *PLOS Biology* 2: 313–17.

Wall, J., and S. Kim (2007). Inconsistencies in Neanderthal genomic DNA sequences. *PLOS Genetics* 3: 1862–66.

Zagorski, N. (2006). Profile of Svante Pääbo. *Proceedings of the National Academy of Sciences USA* 103: 13575–77.

The Causes of Neanderthal Extinction and the Archaeological Record

Barnosky, A. D. (2004). Assessing the causes of Late Pleistocene extinctions on the continents. *Science* 306: 70–75.

Finlayson, C., and J. S. Carrión (2007). Rapid ecological turnover and its impact on Neanderthal and other human populations. *Trends in Ecology and Evolution* 22: 213–22.

Mellars, P. (2004). Neanderthals and the modern human colonization of Europe. *Nature* 432: 461–65.

—— (2006). A new radiocarbon revolution and the dispersal of modern humans in Eurasia. *Nature* 439: 931–35.

SOURCES OF QUOTES

260   *"most barbarous races"*; *"that these remarkable human remains belonged"*; *"it was beyond doubt"*: quoted in Thomas H. Huxley, *Man's Place in Nature*, 150.

264   *"The tree . . . and the associated time scale"*: Cann et al. (1987), 35.

265   *"Homo erectus in Asia was replaced"*: ibid., 36.

266   *"It amounts to killer Africans"*: quoted in R. McKie (2000), *Dawn of Man: The Story of Human Evolution*, New York: Dorling Kindersley, 182.

"fossils are the real evidence": R. Leakey and R. Lewin (1992), *Origins Reconsidered: In Search of What Makes Us Human*, New York: Doubleday, 222.

"paleoanthropologists who ignore": C. B. Stringer and P. Andrews (1988), 1268.

"living genes must have ancestors": A. C. Wilson and R. L. Cann (1992), *Scientific American* 266: 68–73.

267   *"including paleontology, evolutionary biology"*: Higuchi et al. (1984), 284.

270   *"one of these really cool moments of life"*: N. Zagorski (2006), 13576.

"a landmark discovery": T. Lindahl (1997), 2.

272   *"it is not within our power"*: ibid., 3.

Afterword: The Shape of Things to Come

BOOKS

Desmond, A. (1994). *Huxley: From Devil's Disciple to Evolution's High Priest*. Reading, Mass.: Addison-Wesley.

von Humboldt, A. (1859). *Kosmos: A Sketch of a Physical Description of the Universe*. Trans. E. C. Otté. New York: Harper.

ARTICLES

de Beer, G. (1959). "Some Unpublished Letters of Charles Darwin." *Notes and Records of the Royal Society of London* 14: 12–66.

Simpson, G. G. (1960). "The World into which Darwin led us." *Science* 131: 966–74.

WEBSITES

Committee on the Origins and Evolution of Life (2003). "Life in the Universe: An assessment of U.S. and international programs in Astrobiology." Natural Research Council, 47. http://www.nap.edu/openbook/0309084962/html/47.html

"Darwin overview." European Space Agency: Space Science. 17 November 2006. http://www.esa.int/esaSC/120382_index_0_m.html

Jackson, R. "Terrestrial Planet Finder: What is TPF?" NASA Jet Propulsion Laboratory, CalTech. http://planetquest.jpl.nasa.gov/TPF/tpf_what_is.cfm

Meyer, M., Peter Ward, et al. "Rare Earth Debate Part 1: The Hostile Universe." Space.com July 15, 2002. http://www.space.com/science astronomy/rare_earth_1_020715.html

Sodano, Cardinal. "Holy See expresses 'profound sorrow' for Giordano Bruno's death." February 21, 2000. *Daily Catholic* Volume 11, no. 36. http://www.dailycatholic.org/issue/2000Feb/feb21nv2.htm

SOURCE OF QUOTES

278   *"have finally located us on an insignificant mote":* Simpson (1960), 967.

     "a world very different from one conceived": ibid.

     "togetherness and brotherhood with a vengeance": ibid., 970.

     "It is no more true that fruits": ibid.

279   *"If that is a foreordained plan":* ibid.

280   *"countless suns and countless earths all rotating around their suns":* Life in the Universe, 47.

     "profound sorrow": Sodano (2000).

     "The discovery of life on another planet is potentially": Jackson.

281   *"The starry vault and the wide expanse":* Humboldt (1859), 65.

     "by the Creator": See Darwin (1860), *On the Origin of Species*, 2nd ed., London: John Murray, 490.

     "will hereafter be shown to be a part": in de Beer (1959), 59.

283   *"several factors suggest it is common":* M. Meyer (2002).

     "we are over our heads predicting": ibid.

284   *"talk is cheap, exploration and discovery is hard":* A. Knoll e-mail to the author, December 20, 2007.

❦ Index ❧